Bluebells and Nuclear Energy

Bluebells and Nuclear Energy

by Albert B. Reynolds

Cogito Books

Madison, Wisconsin

Published in the United States of America in 1996 by Cogito Books
(an imprint of Medical Physics Publishing)
International Standard Book Number: 0-944838-63-4

Library of Congress Card Number: 96-84323

Cover photograph by Daniel Grogan

Cover design by KC Graphics

Book design and composition by Colophon Typesetting

For information, address the publisher:
Cogito Books
(Medical Physics Publishing)
4513 Vernon Blvd.
Madison, WI 53705

Printed in the United States of America

Contents

Preface

This book is written for several audiences. The first is science teachers in high school, middle and junior high schools, and elementary schools, and their students. A second audience includes those members of the general public who want to learn more than they already know about nuclear energy and radiation. A third is engineers and people who work in the nuclear industry who are not nuclear engineers but want an easy access to a general understanding of nuclear energy and radiation.

Regarding the science teachers, topics related to nuclear energy and radiation are supposed to be taught in different grades in the various sciences. These are often mandated by state departments of education. For the past decade I have been involved, together with my faculty colleagues at the University of Virginia, in offering a summer course in these topics to science teachers. One thing I discovered quite early is that most science teachers are aware that they do not know enough about the subject to do it justice. A second thing I discovered is that the teachers who take our course are extremely eager to learn about nuclear energy and radiation. And learn they do. Their understanding and enthusiasm will go a long way toward dispelling fear of the unknown among their students. There is no better way to appreciate our public school science teachers than to observe them learning new material and sense their eagerness to introduce it into their curricula.

I also discovered that science teachers attempting to teach these topics have a problem. They can only devote a few weeks of their course to nuclear energy and there is a limit to the amount of time they can spend learning about and preparing the material. This problem has guided the design of this book. ***Bluebells and Nuclear Energy*** is not very long, and yet I have tried to cover the most pertinent material that is needed for the various levels and branches of science offered in middle and high schools. Since students are informed through TV and newspapers and ask questions about nuclear topics, the book covers controversial issues related to nuclear energy and radiation in addition to basic technical information. A teacher will not know all of the answers after reading this book; neither do I. Science teachers who take our summer course are aware of this. However, they gain the confidence to know that they know enough that if they don't know the answer to a particular question, it is all right. The subject is complex and one cannot know all the answers.

For the general public, I hope this book will help to dispel some of the fears that so many have about nuclear energy and radiation. The book is short enough that one can learn a lot in a short time. There is some material that may be more detailed and technical than many want, and readers can skip over these parts, but most of this material is useful for the science teachers who form a large part of the book's audience.

There are not many books with general information about nuclear energy for science teachers and the public. I would like to recommend to you two writers, however, who have published excellent books on the subject for the interested science teacher and layperson. The first is Alan Waltar, a former president of the American Nuclear Society, who recently wrote ***America the Powerless: Facing Our Nuclear Energy Dilemma***, Cogito Books, 1995. This book provides an excellent perspective on issues involved in the use of nuclear energy. The second is Bernard Cohen, a professor of physics at the University of Pittsburgh, who has written ***The Nuclear Energy Option: An Alternative for the 90s***, published by Plenum Press in 1990. Dr. Cohen has a remarkable ability to put risks into perspective and to demythologize some commonly held notions about nuclear energy and radiation, and I have incorporated some of his examples into this book.

Some of the material in this book is taken from presentations made by University of Virginia nuclear engineering faculty to the science teachers who have taken the University's summer course referred to earlier. Therefore, I wish to acknowledge the contributions that I received from my faculty colleagues, Jack S. Brenizer, James L. Kelly, W. Reed Johnson, Robert U. Mulder, and Thomas G. Williamson. Detailed reviews and suggestions by Professors John R. Cameron and W. Reed Johnson were especially appreciated. Reviews and figures were contributed by a number of people from Westinghouse, General Electric, ABB Combustion Engineering, and General Atomic. Many figures were provided by the U.S. Department of Energy. I am grateful to Karen K. Harper for her artistry in the preparation of the early drafts of this manuscript, to Richard Montoya for help with the figures, and to Elizabeth A. Seaman for assisting me in final editing. This book has been used as a text for our summer course and many valuable comments and suggestions from teachers have been incorporated.

The use of nuclear energy will become so widespread in the twenty-first century that it is important for more people to understand it better than they do now and especially for the young people in schools to be introduced to the science of nuclear energy and radiation. I hope that this book helps to accomplish these objectives.

ALBERT B. REYNOLDS
PROFESSOR OF NUCLEAR ENGINEERING
UNIVERSITY OF VIRGINIA
1996

Bluebells

Blue is a beautiful color.

Bluebells are my favorite wildflower. Dozens of tiny bells, master crafted by nature, on each plant.

In Tennessee, between Chattanooga and Nashville, there is a small river in the valley just below the mountain where I grew up. In the Elk River is Bluebell Island, a small island that is so absolutely covered with wild bluebells that, in April, the island is a fairyland of blue. When I was a boy, one day in April, the mother of my best friend took my friend and me to Bluebell Island. The sight of all of those bluebells in bloom is one of those indelible memories of my childhood.

Forty years later I learned that they are actually known in plant catalogs as "Virginia" bluebells. Living in Virginia, I planted some Virginia bluebells next to my home and am now enjoying them once again.

In the summer of 1953, after my senior year in college, I experienced another blue color that I will never forget. This was the first time I ever saw a nuclear reactor. The reactor was at the Oak Ridge National Laboratory, and it was under twenty feet of water, which was deep enough that you could look right at it and be perfectly safe. The reactor shone with a wonderfully bright blue glow, characteristic of all "swimming pool" reactors. The blue light surrounds these reactors; it is brightest at the reactor's edge and gradually dies off several feet into the water. The blue is quite like the color of bluebells, pure, deep, like the deepest blue in a perfectly clear sky. I wonder if anyone who sees the blue glow from a nuclear reactor for the first time ever gets over it.

This marvelous blue glow results from a curious source. Gamma rays from the reactor knock out electrons from whatever they hit, such as the hydrogen and oxygen in the water. The more energetic electrons are moving at speeds faster than the speed of light in water, and this situation cannot be sustained. So the electrons slow down to the speed of light in water and in the process radiate electromagnetic energy, the wavelengths of which just happen to overlap the spectrum of visible light. This radiation is called Čerenkov radiation, after the Russian scientist whose experiments led to its explanation. If you ever get the opportunity to see a research reactor, take advantage of it to see this beautiful blue radiation. You cannot see it at a power reactor because the reactor is in a vessel that is in the containment where you cannot go. You may have seen pictures of spent fuel storage pools, however, and they also have the blue glow since the fission products in the spent fuel continue to give off gamma rays long after the fuel is removed from the reactor.

At its source in and around the reactor, the Čerenkov radiation is really a mixture of all colors in the visible spectrum. Such a mixture is white, but this white light appears as a blue glow under water—for the same reason that the sky is blue, a phenomenon explained by Lord Rayleigh a hundred years ago. Short wavelength violets and blues are scattered by molecules (in water or air) more than the longer wavelength reds while, conversely, the reds are transmitted more than the blues. When we look at the sky on a clear day, in any direction other than the sun itself, what we really see is sunlight that has been scattered by air molecules. Since the blues are scattered more than the reds, the sky we see is blue. At sunset the sun's light must travel much farther through the dense part of the atmosphere than during the day so that all of the blues are scattered out of the sun's rays, leaving only the gorgeous reds and oranges. Clouds reflect the reds for even more dramatic sunsets. A similar principle is at work when mountains in the distance are changed to blues and purples. In the case of the Čerenkov radiation, the light directly in line with the high-intensity source at the reactor still contains enough blue to look white, but surrounding the reactor we see only the scattered blues, which gives us the beautiful blue glow.

I want to keep the bluebells growing as much as the next person does. And the sky blue and the mountains purple and the waves of grain amber. Clean energy is a way to keep this happening. Conservation and renewable sources of energy will help. And nuclear energy can lead the way.

1

Introduction

People who work around nuclear reactors do not fear them. Nor do they fear radiation. None of them do. They understand and respect reactors and radiation. Fear is simply not an emotion that they experience with respect to radiation and the routine generation of electricity from nuclear energy. On the other hand, some level of fear of radiation and nuclear energy appears to be almost universal among the general public. Maybe this book will help dispel some of this fear. This is certainly one of its objectives.

There is little doubt that nuclear energy will be one of the world's dominant sources of electricity throughout the twenty-first century. As the third world strives for energy and a reasonable standard of living, the need for additional electricity generation will become enormous. As the developed world seeks to curb the addition of gases that contribute to the greenhouse effect, like carbon dioxide (CO_2), which is produced with the burning of all fossil fuels, we will inevitably turn increasingly to nuclear energy, which is both economical and produces no CO_2. Fear of nuclear energy will disappear with its increasing use in the next century just as surely as the widespread fear of electricity itself vanished in the early part of the twentieth century. In my view, it is only a matter of time before the people most concerned with the world's environment recognize the value of nuclear energy when the combination of environmental and economic considerations are weighed.

An understanding of radiation is important to the understanding and appreciation of nuclear energy. An introduction to radiation is provided in chapter 2. Some of the widespread uses of radiation are discussed in chapter 3. A discussion of fission and the use of nuclear energy for the generation of electricity begins in chapter 4. Also in chapter 4 I have included some history of radiation, fission, and nuclear power.

No one can say exactly when new growth in nuclear electricity will begin in the United States, but, in time, it will happen. At present natural gas is inexpensive, and in the present cost-competitive atmosphere, which favors short-term solutions, natural gas will be the primary source of new electricity during the 1990s. Many utility planners are convinced, however, that natural gas prices will eventually rise so that electricity from nuclear plants built for operation in the next century will be cheaper than gas. Coal accounts for over half of the electricity generated in the United States, and efforts are underway to make coal environmentally more acceptable than it is now, including the gasification of coal, so that electricity from coal may remain competitive with nuclear electricity.

For the long run, nuclear energy will clearly be necessary in both the United States and throughout the world. Projections based on expected population growth, per capita needs for electricity, and huge improvements in energy efficiency indicate that by the middle of the 21st century the world will need three or four times as much electricity as we have today—a staggering increase. To accomplish this economically and acceptably from an environmental standpoint will require the greatly expanded use of nuclear energy. I am talking here about nuclear energy from fission; the other kind, fusion, cannot be made competitive commercially before the middle of the next century, if ever. Renewable sources of energy, like biomass and solar, neither of which are close to being economical yet, will also be needed, but they cannot begin to fill the gap. Although natural gas produces less gas that contributes to the greenhouse effect than coal and is currently cheap, there is no way that natural gas can supply the growing needs of the next century. Massive amounts of nuclear energy will be needed.

There are several reasons for the renewed growth of nuclear energy in the United States. One is that a new generation of nuclear power plants has been designed in the 1990s for construction after 2000 that satisfies the fundamental requirement that they be economically competitive with other methods of generating electricity. A second reason is that the new designs satisfy the level of safety demanded by modern society. A third reason is that nuclear energy satisfies the need for an environmentally benign source of electricity and produces no gases that contribute to the greenhouse effect. These features coincide with the expected continued increase in the demand for electricity in the United States, even accounting for conservation.

The attractiveness of nuclear energy is eminently apparent to those who understand it. Fear of radiation is widespread among Americans, and this fear has blocked the understanding of the safety and value of this remarkable energy source. The anti-nuclear movement in the United States is fueled by this fear. Understanding radiation and recognizing how small the amount of radiation that the commercial nuclear energy enterprise adds to the environment can eliminate this fear.

It's not that things cannot go wrong with a nuclear power plant. But it is just short of impossible to endanger the public with the types of plants presently operating in the United States, Western Europe, and Japan, and even less likely with the new designs. One can indeed imagine sequences of several events, all of which must occur one after the other, that might cause damage outside of the plant. A water pipe might break or the water system spring a leak. All of the backup systems might fail. Electricity from the utility's power grid might be cut off. After that, all of several alternate electric sources might fail to start. Or operators might make a series of mistakes following a pipe break or leak. Even if all these things happened at the same time, it is still unlikely that the safety of anyone outside the plant would be affected. Only once in the approaching half-century history of western commercial use of nuclear energy has more than one of these events in the above sequences occurred at the same time.

This was at Three Mile Island, where two of the above events occurred, and in this case there was no damage at all to the health of the surrounding population or the plant personnel. Only the reactor was damaged. (I do not include Chernobyl here since that reactor design could never have been licensed for operation in any western country.)

Now what if nuclear plants were designed so that if all of the above events were to occur and the operators did absolutely nothing (and I mean literally nothing), the reactor would not be harmed at all? Some of the new designs of the 1990s are such reactors. Other new designs improve on the safety of present plants in other ways.

How reactors like this can be designed is described in this book. And these reactors are economically competitive with coal and natural gas and with the nuclear plants that began operation before the 1990s.

Nuclear energy is used mostly for the generation of electricity. This, together with ship propulsion, is about the only widespread use of nuclear energy for power to this point. It has been and is being further developed for propulsion in space; in fact, it is the only means available for sending spacecraft on long missions, say to Mars and beyond, and for supporting significant manufacturing operations and personnel colonies in space. It will also be widely used someday for process heating, which is the term used to describe the energy driving chemical reactions for the vast chemical industry. It is now being used for space heating in Russia; this means that steam is produced by a reactor and transported through pipes to heat an entire town or section of a city to heat homes and other buildings.

In this book, however, we want to talk about electricity from the atom. That's where the big action is. And much of it is happening in the United States. Nuclear power was big in the United States in the late 1960s and the seventies. In the 1980s, nuclear energy became the dominant source of electric power in France and a large source in the rest of Europe, Japan, and Canada. New construction is continuing in Japan, where future plans for nuclear growth are especially ambitious. In Europe the pace of new construction has slowed as the need for additional electricity has slowed. The next big market is the "Pacific Rim," composed of China and Taiwan, Japan, Korea, and Indonesia. Many expect that eventually the United States will become a large market again.

While Europe and Japan have been building the types of reactors in the eighties that the United States built in the seventies, the United States has been designing the next generation of reactors. The Europeans and the Japanese have been noticing—and are doing everything they can to follow our lead. This has resulted in the internationalization of the nuclear power industry, much like what is happening in business everywhere. The U.S. companies that have been in the forefront of nuclear development are Westinghouse, General Electric, Combustion Engineering, Babcock & Wilcox, and General Atomic. Look what happened to these companies in the late 1980s and early 1990s. Combustion Engineering was bought out, lock, stock and barrel, by ASEA Brown Boveri, the largest electrical supply company in the world, which itself resulted from the merger of Sweden's ASEA Atom and Switzerland's Brown Boveri. Babcock & Wilcox's commercial nuclear business was bought out by France's nuclear giant, Framatome. The nuclear components of General Electric and Westinghouse have not formally combined with European or Japanese companies. However, General Electric and the Japanese companies Hitachi, Toshiba, and Tokyo Electric have jointly designed an advanced nuclear plant, and they are currently building two of these plants in Japan for operation in 1996 and 1997. Westinghouse is designing advanced reactors with Mitsubishi. France (Framatome) and Germany (Siemens) have teamed up to design a new "European" reactor. This is probably just the beginning of the internationalization of nuclear energy.

While nuclear engineers and the U.S. Government's safety regulatory agency, the Nuclear Regulatory Commission (NRC), agree that the 1970 style nuclear plants are safe enough, many in U.S. society have not agreed. These plants do require some action by reactor operators and some positive response from equipment in the event of some accidents, and for many Americans this has not been acceptable. For some accidents this equipment response requires a source of electricity, for example to run pumps. So, while the Europeans and the Japanese were ordering U.S. 1970 style nuclear plants during the 1980s and U.S. nuclear construction was brought to a standstill by a combination of fear and economic conditions, a whole new generation of safer nuclear plants was being designed in the United States. U.S. engineers designed some plants that would not require operator action—for any accident. I mean no operator action at all—zip, zilch. They also designed plants so that the equipment to handle an accident requires no electricity other than the early use of stored electricity from batteries. For some plants this means the operation of some valves that operate automatically when electricity is lost or rely on batteries for their electricity. For other designs neither valves nor any other pieces of equipment with mechanical parts have to work at all. All that is required is for water to flow downhill, like when it rains, and for hot air to rise, like what happens every time you build a fire in your fireplace.

Actually I'll have to back off, or qualify my credit to U.S. engineers somewhat in the last paragraph. The first of these amazing reactor designs was really invented in Sweden. But four such reactor designs have been developed in the United States since that time. These designs are referred to as ***passive*** plants and are described in chapter 7.

In addition to the ultrasafe designs referred to above, U.S. designers, in some cases working with the Japanese, have simplified present reactor designs, making them significantly more reliable and further reducing the already small chance of accidents, while still keeping a source of electricity to run pumps in the event of an accident. Japan is building two such plants designed by General Electric. These designs are referred to as ***evolutionary*** plants.

In order to understand why the new designs are so safe, however, it is necessary to understand how a nuclear plant works and how the present plants assure safety. Therefore, the general operation of a nuclear power plant is described in chapter 5, and the safety of the presently operating nuclear power plants is described in chapter 6. This is followed by a description of the automatic safety operation of the new plants in chapter 7.

One of the most misunderstood, and therefore most fearful, of all nuclear energy issues is radioactive waste. Disposal of high-level nuclear wastes is discussed in chapter 8.

Let me say a little bit at the outset about weapons and defense. This book is about commercial uses of nuclear energy. It is not about weapons and defense other than a few words in the last chapter. Nuclear defense has been shrouded in secrecy since the 1940s, with no regulation by any public body. The result is that nuclear matters related to the nation's defense have been conducted with insufficient regard for the environment. On the other hand, commercial nuclear power has been scrutinized thoroughly by the public from the beginning, that is, from the middle 1950s. The difference is that the safety record of commercial nuclear power is excellent, while environmental problems associated with some defense installations have been deplorable. The anti-nuclear movement often blurs the distinction between defense and commercial nuclear power whereas the histories of the two have been totally different. Perhaps it is asking too much for the media to help formulate this distinction in the minds of the general public. The fact is, however, that transgressions at some of the defense installations have no bearing whatever on the ability of commercial nuclear energy to be handled safely.

SOURCES OF ELECTRICITY

It is useful to know where our electricity comes from. At present the dominant source of electricity in the United States is coal, with nuclear energy being second. The breakdown according to source in the United States in 1995 was as follows:

Coal	51%
Nuclear	20%
Natural Gas	15%
Hydroelectric	9%
Renewable Sources	3%
Oil	2%

TABLE 1.1 Electricity Generated by Nuclear Energy in 1994 Worldwide

Country	*Percentage of Each Country's Total Electricity Produced by Nuclear Energy in 1994*
Lithuania	76
France	75
Belgium	56
Sweden	51
Slovak Republic	49
Bulgaria	46
Hungary	44
Slovenia	38
Switzerland	37
South Korea	35
Spain	35
Ukraine	34
Taiwan	32
Japan	31
Finland	30
Germany	29
Czech Republic	28
United Kingdom	26
United States	22
Canada	19
Argentina	14
Russia	11

Renewable sources of energy include solar, wind, biomass, and geothermal. Be wary of the fact that many who like renewable sources and don't like nuclear energy frequently lump hydroelectric together with solar and wind and call them all renewables in order to make renewables sound like a large current contributor. Actually, wind, direct solar, and biomass will likely contribute to some extent in the future. However, most observers think these renewable sources of energy will not be able to compete economically with nuclear energy on a large scale for a long time, and even then mostly only for special uses. I discuss this further at the end of this chapter.

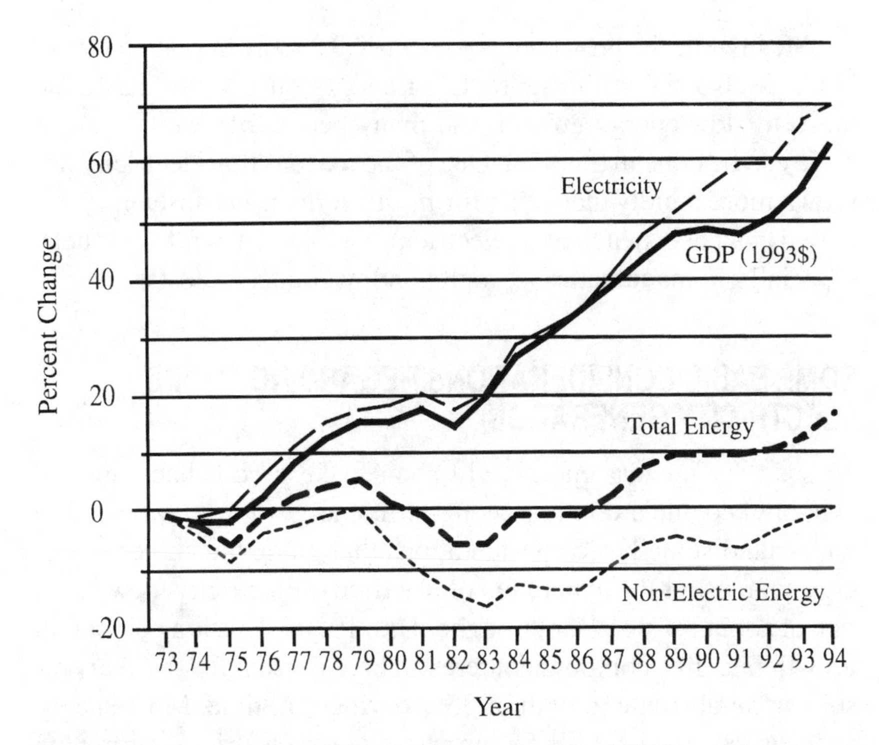

Figure 1.1 Comparison of U.S. electricity, energy, and GDP growth rates. COURTESY OF NUCLEAR ENERGY INSTITUTE.

The country with the highest percentage of nuclear electricity (other than small Lithuania) is France; 75% of their electricity is generated by nuclear energy. For the world, nuclear energy's share of electricity generation is about 17%. The fraction of electricity that was generated by nuclear energy in 1994 in each country is given in table 1.1, as reported by the United Nations' International Atomic Energy Agency.

Presently in the United States the rate of increase in the use of electricity almost exactly parallels the rate of growth of the gross domestic product (GDP). The growth of electricity and GDP are compared in figure 1.1. While there is no apparent fundamental reason why the two have to advance in such lockstep (which means that it won't necessarily continue), the similarity in performance of the two for the past two decades is phenomenal.

Meanwhile the growth in non-electrical energy has fallen far behind, as also shown in figure 1.1. Energy conservation has done much to slow energy growth, and many people believe that much more can be done in this area. One of the reasons that electricity has grown more rapidly than other forms of energy is because many energy users have switched to electricity and away from fossil fuels, especially in manufacturing, to obtain lower energy costs.

SOME BASIC CONSIDERATIONS REGARDING ELECTRICITY GENERATION

To see why nuclear energy will soon make a comeback and inevitably become a dominant long-term energy source, we need to understand some basic considerations that guide the choice of energy sources for the generation of electricity. Electricity grew at the rate of about 3% per year throughout the 1980s. This is a compound growth rate, like compound interest. Three percent for ten years results in an absolute growth of 35% during that time. Most energy analysts expect about a 1.5% growth rate for electricity during the late 1990s and early 2000s, slightly less than the GDP. This would lead to a 25% increase in electricity use between 1995 and 2010.

The capacity of the U.S. electrical grid in 1995 was about 800 gigawatts (GW) of power (where ***power*** is the ***rate*** of using energy, i.e., energy per unit of time, and a gigawatt is 1 billion watts). Of this, about 100 GW were produced by nuclear energy. Gigawatt is a good metric unit, and the gig is officially pronounced like jig, as in dancing a jig (though you will hear the initial g often pronounced as a hard g). ***Energy*** is power times time, hence the energy you are charged for in your electricity bill is given in kilowatt hours. A kilowatt (kW) is 1000 watts. A megawatt (MW), which we shall also use, is 1 million watts. A 25% increase in capacity between 1995 and 2010 would mean that 200 GW of electrical generating capacity would be added to the grid by 2010. Further additions would be needed in the decades after that.

Another factor in the equation is reserve, or margin, in the system. There is much variation in electricity demand between seasons and between the peak and the average during a given season. Electricity demand peaks on the hottest summer days and the coldest winter days. Utilities like to keep a margin of about 17% between capacity and peak demand since some of their power plants are often down for repairs or, in the case of nuclear plants, for refueling. During the 1980s the margin was generally greater that 17% so that few large power plants were ordered—i.e., no nuclear plants and hardly any coal plants. However, as utilities entered the 1990s, the margin for many utilities was right at 17%, so that these utilities needed to add capacity.

Broadly speaking, there are two kinds of capacity that can be added. The first is ***base-load*** capacity and the second is ***peak-load*** capacity. Base-load plants run all the time. Peaking plants are up and down to account for the variation in daily and seasonal use of electricity. You can reason that the plants with the cheapest fuel would be the ones utilities would want to use for base loading. These also happen to be the ones that cost the most to construct. On the other hand, plants that are cheap to build, even if the fuel is expensive, would be ideal for peak loading since they have to be run only part of the time.

A potentially large new use of base-load electricity is the electric vehicle. Environmental considerations have led California to mandate the introduction of electric cars in that state by 1998, and the auto companies are moving closer to economical designs. The batteries that power these cars will generally be recharged at night using base-load electricity when load requirements are low, an ideal use for nuclear energy.

The cost of building a plant is called the ***capital cost***; the cost for the fuel is called the ***fuel cost***. Nuclear plants are the most expensive to build and, hence, have the highest capital cost, but they operate with the lowest fuel costs. For this reason they are used predominantly for base loading (although in France, where 75% of the electricity comes from nuclear energy, they are also used for some peak loading). A third cost is the ***operation and maintenance cost***. In recent years operating and maintenance costs have been high for nuclear plants, and great effort is currently underway to reduce these costs. Large coal plants are also used mostly for base loading. Another feature of the base-load plants is that they take longer to build than peak-load plants.

The highest-cost fuels are oil and natural gas, but gas turbines and oil-fired plants are the least expensive to build; hence they are used for peaking. In the volatile field of energy, there was a time when oil was so cheap that coal plants were converted to oil. When oil prices rose, they were converted back to coal. For a while it was illegal to use natural gas for a large base-load plant because we thought there was a shortage of gas. As I write this, there is no shortage of gas and the price is right, so once again gas is back in favor, even in some cases for base-load plants and especially for cheap peaking gas-turbine plants. Who knows what the price situation will be by the time you read this. As we entered the 1990s, base-load capacity was still adequate in most of the country, but peak-load additions were needed. Therefore, utilities have been adding mostly small plants in the 1990s, even using relatively expensive natural gas. However, for large base-load additions, nuclear is competitive with coal and gas in most parts of the country. (See the end of chapter 7 for more on the relative costs for the various fuels.) When will we need new base-load capacity? Projections indicate that large additions of base-load plants will be needed shortly after the year 2000, and this means that new base-load capacity will have to be ordered in the late 1990s.

Still another factor is the use of ***cogeneration*** plants. Cogeneration means the use of one plant for two purposes—for process heat and for the generation of electricity. The high-temperature steam produced is used as a heat source for chemical manufacturing or other process and the lower grade steam is used to generate electricity. Cogeneration plants are often built and operated by chemical companies, which need the process heat; the electricity is used by the plant, with any excess being sold to a local utility. Cogeneration accounted for a sizable fraction of new electricity additions during the 1980s.

LONG-RANGE THINKING: WORLD ELECTRICITY, ENERGY SOURCES, CONSERVATION, AND GLOBAL WARMING

Accurate energy projections are very difficult to make; I am not in the business of making such projections, though I am an avid reader of those who do so. I am aware of the advice of the widely quoted New York Yankee catcher and manager, Yogi Berra, who said, "You should never trust a man who makes predictions, especially about the future." Despite his advice and the uncertainty of projections, it is still important for people to look into the future to help guide our planning.

The results of long-range projections of the world's electricity needs are startling. A most interesting view was provided in 1990 by long-time energy leader Chauncey Starr and his colleague Milton Searle in the book ***Conversations about Electricity and the Future***. They were concerned with electricity needs, probable electricity sources, and CO_2 production from electricity generation until the year 2060. First they reported on population trends, which are uncertain but which, nevertheless, drive energy demand. They then observed present trends in per capita electricity growth. To account for energy conservation, they assumed that in the future electricity use would be reduced to two-thirds of its present per capita growth trend and that other direct energy uses would be reduced to one-half of the present trend. (Recall that present trends have already incorporated the conservation measures that have been taken since the 1973 oil embargo.) For the year 2060 the results show a five-fold increase in the world's demand for electricity over 1986 levels. They project a 3.5-fold increase in the developed countries and a 10-fold increase in the less developed countries.

This is a huge amount of growth. Part of the reason for the growth is that electricity usage is growing much faster than energy usage in general, and this is expected to continue. The rest of the reason is the projected growth in world population by 2060 by a factor of 1.9 (1.5 in the developed countries and 2.0 in the less developed countries).

They then discuss where all of this electricity might come from. Fossil fuel will be one competitor, as it is today, but direct energy uses such as transportation and space heating will make great demands on oil and natural gas. Starr and Searle recognized the need for conservation in their model, plus the need to develop renewable energy sources to the extent possible. They estimated maximum amounts for contributions from solar and biomass. Their upper limit estimate for the solar electricity contribution assumes that solar contributes all of the peak load during the daylight hours, which accounts for about 13% of the total electricity generated. Their upper limit for biomass is governed by water and land availability and comes to 23% of the total. Thus, by their reckoning, solar plus biomass might account for 36% of the world's electricity by 2060. Although neither solar nor biomass is yet a proven economic competitor, except for the generation of electricity from trash, and combined they contribute only about 3% of today's U.S. electricity supply, Starr and Searle assumed that somehow these sources could compete and make a huge contribution by 2060.

Now what does this leave for the other three potential sources—nuclear, fossil, and hydroelectric power? Fossil fuels will continue to be used for many purposes in addition to generating electricity. Burning fossil fuels produces CO_2 gas, which leads to the greenhouse effect. (An expanded discussion of the greenhouse effect appears in chapter 9.) Unlike nuclear energy and the renewable sources of energy, the waste issue for fossil fuels with regard to CO_2 is unresolvable. Therefore, there will eventually be an effort to limit the use of fossil fuels for electricity. Hydroelectric power is limited by available sites. If fossil fuels and hydroelectric power are held at the 1986 levels, they will contribute only 17% of the world's electricity in 2060, according to Starr and Searle's projections, and nuclear energy would have to generate 47%. The absolute amount of nuclear electricity generated would be 14 times as much as nuclear energy produced in 1986. If fossil fuels and hydroelectric power increased so that combined they produced as much electricity as nuclear energy in 2060, then fossil fuels and hydroelectric power would produce 32% of all electricity, and nuclear energy would produce 32%.

In this case the growth in nuclear energy over the 1986 levels would be by a factor of 10. This is a huge increase.

Remember that nuclear energy is economically competitive with gas and coal in most parts of the developed world today and is more economical than oil except for peak loads. Renewable sources of energy are not yet economically competitive, and many question whether solar power will ever be except for special applications. What all of this means is that nuclear energy will almost certainly be one of the major sources of electricity in the twenty-first century.

It is unfortunate that many who fear nuclear energy argue that renewable sources of energy combined with conservation can provide all of our electricity in the future. This is an illusion. The illusion gains credence among an unwitting public when sensational media presentations and TV ads made to appeal to people especially concerned about the environment create the impression that competitive electricity from renewable energy sources is just around the corner. This is not so except in very special cases. It is the hope of most people involved in the nuclear energy enterprise that renewable sources can contribute in a big way someday in the future since it is difficult to imagine how nuclear energy and fossil fuels can do it all. It is not an either/or question of nuclear energy versus renewable sources. Both are needed. We ought to develop conservation and renewable sources to whatever extent possible. Even if we do, however, there will be a huge gap that must be filled with nuclear energy. The widely proclaimed political position that renewable energy sources can do it without nuclear is an irresponsible hoax.

A final question to be considered before we talk about how a nuclear power plant works is whether there will be enough uranium to fuel all of the nuclear plants that will be needed in the centuries to come. The answer is yes there is. There would not be enough uranium for our present type of reactors. However, there is another type of reactor, called a breeder reactor, for which there is enough uranium to last forever. How a breeder reactor works is discussed near the end of chapter 5 after enough information has been introduced to understand this kind of reactor.

2

Nuclear Radiation

To a great extent, the widespread fear of radiation discussed in chapter 1 is, it seems to me, based on unfamiliarity. People who understand radiation do not fear it. They respect it and are careful to prevent exposure to radiation that could be harmful, but scientists and engineers accomplish this routinely all the time.

For our purposes we are interested in ***nuclear radiation***, i.e., radiation that emanates from the nucleus. The primary types of nuclear radiation are alpha particles, beta particles, gamma rays, and neutrons. Others are protons and positrons, but they play a minor role in relation to radiation from nuclear reactors and radioactive materials. These are all forms of ***ionizing radiation*** because when they interact with matter they produce ***ions***, which are charged particles. The production of ions by neutrons is a secondary process since the neutron does not produce ions directly in a collision with a nucleus. Non-nuclear forms of ionizing radiation are x-rays and high-energy ultraviolet radiation, like the harmful radiation from the sun. Non-ionizing radiation includes thermal radiation, low-energy ultraviolet radiation, visible light, infrared radiation, microwaves, radio waves, and others. Ionizing radiations are of particular interest to us both because they have many extremely beneficial uses and because large amounts of ionizing radiation can have detrimental health effects. Unlike light and thermal radiation, low-level ionizing radiation cannot be detected by a person's normal sensing mechanisms.

Before discussing the various types of nuclear radiation, it is useful to review some basic concepts about atoms and nuclei.

ATOMS AND NUCLEI

We will keep what we need to know about atomic and nuclear chemistry and physics to a minimum. Recall that an atom is made up of a nucleus and its surrounding electrons. Modern physics talks about an electron cloud, or electron shells, but for our purposes we don't need to distinguish between this and the old familiar model of electrons orbiting around the nucleus—like planets around the sun. A helium atom is shown in figure 2.1. The nucleus contains protons and neutrons, both of which are about the same mass. (Note that I use "mass" instead of "weight." Scientists are for the most part careful people who distinguish between the two, but we do not have to bother with the difference here. You can consider them the same for our purposes.) Protons and neutrons have about the same mass and are about 1840 times more massive than electrons.

The number of protons in an atom is the ***atomic number***, denoted by Z. The number of protons plus neutrons is the ***mass number***, denoted by A. These numbers are shown for helium in figure 2.1 in the way that elements are often written in scientific shorthand, that is, ${}^{4}_{2}He$. Any individual atomic species (i.e., specific A and Z) is called a nuclide.

Electric charges are associated with the proton and the electron. Each proton carries one positive charge; each electron carries one negative charge. The neutron carries no charge; it is neutral. Hence its name. Quite logical, as scientists like to be. Each atom is also neutral electrically, which means that the numbers of protons and electrons in an atom are the same. Of course, you remember all this from high school chemistry. On the other hand, many of you who aren't science teachers might have forgotten; it's been a long time.

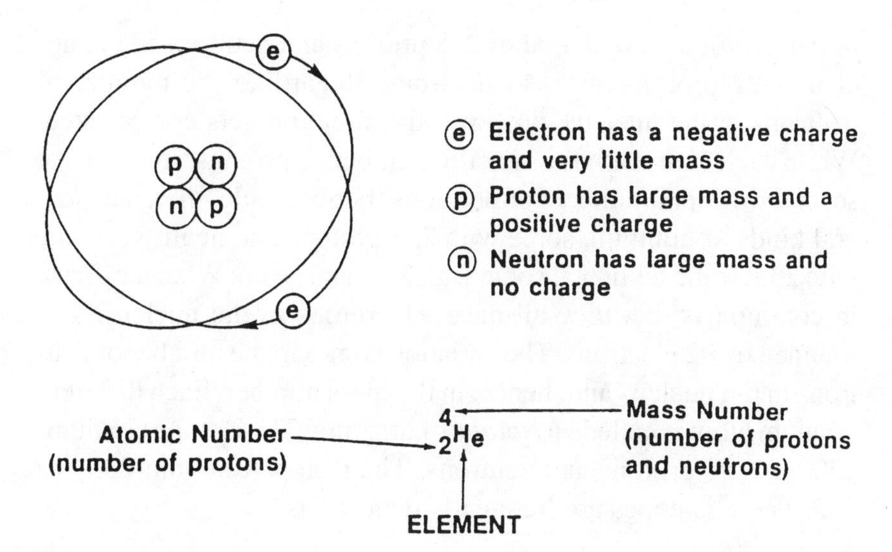

Figure 2.1 The helium atom and nucleus. COURTESY OF PENNSYLVANIA POWER & LIGHT COMPANY.

The mass of different atoms depends on the number of protons and neutrons in the nucleus. A hydrogen atom has only a single proton in its nucleus (and a single electron revolving around it); thus it has an ***atomic weight*** of one, or about one. (It's not weight at all; it is mass, but scientists got on the wrong track a long time ago and called it atomic weight. We're trying to correct it now. The new official name is ***molar mass***, but most of you probably learned it as atomic or molecular weight.) I say the atomic weight of the hydrogen atom is about one; the mass of the proton that forms the nucleus of the hydrogen atom is actually 1.007 28 in what we call ***atomic mass units*** (amu). One amu is equivalent to $1.660\ 56 \times 10^{-27}$ kg. (For those interested, this is the inverse of Avogadro's number, 6.022×10^{26} molecules per kilogram mole.)

Uranium is an ***element***, like hydrogen, iron, etc. Each element has a given number of protons in its nucleus. Every uranium nucleus has 92 protons; hence, the atomic number of uranium is 92. Most uranium atoms have a total of 238 protons and neutrons in the nucleus—92 protons and 146 neutrons. Regarding the number of neutrons in the nucleus, however, the situation gets complicated. While each element has a specific number of protons, it may have several different numbers of neutrons. For example, there are several kinds of uranium, some with 238 protons and neutrons, some with 235, some with 233, some with 239, and so on. What they have in common is that they all have 92 protons in the nucleus, surrounded by 92 electrons. The variation comes in the number of neutrons in the nucleus and, hence, in the mass number. Each different uranium atom is called an ***isotope*** of uranium. The isotope uranium-238 has 238 protons and neutrons. The isotope uranium-235 has 235. These isotopes are frequently denoted as

$$^{238}_{92}U \text{ and } ^{235}_{92}U.$$

Another abbreviated notation for isotopes, which is convenient for writing in prose since it doesn't require subscripts, is U-238 and U-235.

Hydrogen has isotopes too. Ordinary hydrogen has one proton in its nucleus. Another hydrogen has one proton and one neutron. This is called ***deuterium***, or sometimes, colloquially, heavy hydrogen; its mass number is 2, and it is often abbreviated D. Then there is hydrogen with a mass number of 3. It has 1 proton and 2 neutrons and is called ***tritium***.

Most nuclides found in nature are ***stable*** nuclides. They will remain the way they now are forever as long as people don't tinker with them or some cosmic ray or other natural particle doesn't strike them. Some nuclides are ***unstable***. These include some found in nature, some produced in nuclear reactions in machines or devices made by people, and some produced by nuclear reactions that occur in nature. Unstable nuclides change into stable nuclides by radioactive decay. Many elements have more than one stable isotope. Ordinary hydrogen and deuterium are stable isotopes of hydrogen, while tritium is unstable. All uranium isotopes are unstable, but some of them decay so slowly that, fortunately, they are still around in nature.

NUCLEAR REACTIONS, RADIOACTIVITY, AND TYPES OF RADIATION

Types of Nuclear Radiation

The various types of nuclear radiation are described in table 2.1. Most have a charge. This charge is the charge of an electron, which is 1.6×10^{-19} coulomb. All but one have mass. The mass of nuclear particles is generally given in atomic mass units (amu).

The ***LET*** in the table stands for ***Linear Energy Transfer***. This is the amount of energy transferred to a material per unit distance that a particle or wave of ionizing radiation travels in the material. If radiation has a high LET, it deposits its energy in a short distance, and, consequently, causes a lot of damage over a short distance. Radiation with a low LET causes less damage per unit distance since its energy is transferred over longer distances. (To be more exact, as scientists like to be, the gamma ray and neutron do not actually have LETs themselves. The low and high LETs associated with them are the result of electrons produced by gamma ray interactions and heavier particles produced by neutron interactions.)

Alpha particles can be stopped by a thin sheet of paper or an inch or two of air. Beta particles can penetrate paper, but not a thin sheet of metal. Gamma rays and neutrons can penetrate a thick piece of material.

TABLE 2.1 Types of Nuclear Radiation

Type	*Symbol*	*Mass* (amu)	*Charge* (electron charges)	*Relative LET**
Alpha Particle	α	4.0026	+2	High
Beta Particle	β (or β^-)	1/1823	−1	Low
Gamma Ray	γ	0	0	Low*
Neutron	n	1.0087	0	High*
Proton	p	1.0073	+1	High
Positron	β^+	1/1823	+1	Low

*See qualification in the paragraph below on LET.

Somewhat colloquially, I am distinguishing in table 2.1 between a beta particle and a positron. To be more accurate, a positron is a positive beta particle and an electron is a negative beta particle. Negative beta particles are what we are concerned with in radiation from nuclear reactors, so when I use the term beta particle I am assuming the negative beta particle.

Radioactivity

A material is ***radioactive*** if its nucleus is unstable. The nucleus of a radioactive material changes spontaneously by one of several processes, including emission of an alpha particle, a beta particle or a positron, and/or a gamma ray. This process is called radioactive decay, or ***radioactivity***—the name given to it by Marie Curie. Radioactive nuclides are called ***radionuclides***. Radioactive isotopes of a particular element are called ***radioisotopes*** of that element. When an element decays by alpha, beta, or positron emission, the nucleus changes into a new element. The original nucleus is called the ***parent***, or ***precursor***. The new element is called a ***daughter***, or ***progeny***. The progeny may be stable or unstable. If unstable, it too will undergo radioactive decay.

Another radioactive decay process, called an ***isomeric transition***, occurs when a gamma ray is released from an excited nucleus without creating a new nuclide. In this case, the two states of the nucleus, before and after the emission of the gamma ray, are called ***isomers*** of the nuclide. The excited nuclide has an "m" beside its atomic number, referring to its metastable, or excited, state. An example is technetium-99m, or ^{99m}Tc, which decays to ^{99}Tc.

Additional radioactive decay processes are possible, for example, electron capture and internal conversion, but they are not important for our general understanding of radiation and won't be described here.

Materials are unstable, and hence radioactive, whenever they have "too many" neutrons or protons in the nucleus. Except for hydrogen-1 and helium-3, stable nuclides have either an equal number of neutrons and protons or more neutrons than protons. Bismuth is the most massive element for which a stable isotope exists. The single stable bismuth isotope, Bi-209, has 83 protons and

126 neutrons. For most elements two or more stable isotopes exist. For elements lighter than bismuth, if an unstable nuclide has too many neutrons, it decays by emitting a beta particle. If the nuclide has too many protons, it decays either by emitting a positron or by electron capture. Alpha decay is limited to the heavy elements.

Units of Radioactivity

The SI (Systeme International) unit for radioactivity is the ***becquerel*** (Bq), which refers to one radioactive disintegration per second. The becquerel is used in two senses. First, it is the ***quantity*** of a radioactive nuclide in which the rate of decay is one disintegration per second. Second, it is a measure of ***activity***, which is the rate at which a radioactive material decays, in disintegrations per second, or becquerels. An older unit (which is still widely used) is the ***curie*** (Ci), which is the quantity of a radioactive nuclide in which the number of disintegrations per second is 3.7×10^{10}. Thus, 1 Ci = 3.7×10^{10} becquerels. A curie is also about equal to the decay rate of 1 gram of radium-226, which was the earlier historical definition for this somewhat odd unit.

Energies in Radioactive Decay and Nuclear Reactions

Radioactive decay is a nuclear reaction. In this reaction the sum of the masses of the decay products is different from the mass of the original radioactive nucleus. The difference in mass has been converted into energy (for example, as a gamma ray or kinetic energy of the products) according to the famous relationship discovered by Einstein,

$$E = mc^2.$$

Here E is the energy, m is the mass, and c is the velocity of light (3.0×10^8 metres/second).*

The unit of energy that we generally use to describe nuclear reactions is the MeV, which stands for million electron volts. Thus

*Note my spelling of metre. This is not a mistake; this French-style spelling of metre is the official spelling in the International System of Units, abbreviated SI from the French name, Le System International d'Units.

1 MeV = 10^6 eV. One MeV is a small amount of energy, equivalent to 1.602×10^{-13} joules. (A 100 watt light bulb uses energy at the rate of 6×10^{14} MeV per second.) However, when the MeV per reaction is multiplied by the number of reactions in a material, total reaction energies can become quite large. The energies of alpha and beta particles and gamma rays from radioactive decay are generally in the MeV range.

Alpha Particles

Alpha particles are produced from the natural radioactive decay of heavy elements, such as uranium and plutonium, or from nuclear reactions. They are composed of two protons and two neutrons and thus are equivalent to the nucleus of the helium atom (which was demonstrated by Rutherford ninety years ago). The two protons give the alpha particle a positive charge of two units. An alpha particle is massive compared to other forms of radiation.

An alpha particle can travel only a very short distance in a material before it is stopped; as stated earlier, a sheet of paper can stop an alpha particle. Alpha particles can travel only about one inch in air and 0.001 inch in human tissue. A person's skin is constantly being regenerated so that on the surface of a person's skin there is always a dead layer. An alpha particle cannot penetrate this layer; hence it is not an external hazard to the body. It becomes a hazard only if it is ingested into the body. A famous example is the radium dial painters who, while painting radium onto the dials of watches, would lick the tip of the paint brush to obtain a fine point. In the process some of them absorbed enough radium in their bones to die of bone cancer.

Alpha particles generally have energies in the MeV range. For example, the alpha particle from the decay of radium-226 has an energy of 5 MeV. This energy is in the form of kinetic energy ($\frac{1}{2} mv^2$, where m is the mass of the alpha particle, and v is its velocity). When the alpha particle is stopped in a material, it gives up its energy to the material, thereby causing ionization and damage to the material.

When an element decays by alpha emission, it changes into another element. Recall that an element is identified by the number of protons in its nucleus, which defines its atomic number, Z. In alpha decay a new element is formed with an atomic number lower by two than the atom that decayed. The mass number decreases by four since two neutrons are lost in addition to the two protons. Thus, for example, when uranium-238 ($^{238}_{92}U$) decays by alpha emission, the resulting nuclide is thorium-234 ($^{234}_{90}Th$). When plutonium-239 ($^{239}_{94}Pu$) decays by alpha emission, the daughter is uranium-235 ($^{235}_{92}U$). When radon-222 ($^{222}_{86}Rn$) decays by alpha emission, the daughter is polonium-218 ($^{218}_{84}Po$).

Beta Particles

Beta particles are indistinguishable from electrons, at least negative beta particles are. The only difference between a beta particle and an electron is that a beta particle comes from the nucleus—it's really just semantics since they are identical. A beta particle actually comes from the decay of a neutron! A neutron decays by ejecting a beta particle, and in the process the neutron turns into a proton. As seen in table 2.1, a beta particle is a very light particle. Like an electron, its mass is a tiny fraction of the mass of a neutron or proton, about 1/1840 of the neutron or proton mass.

A beta particle can travel up to about fifteen feet in air and half an inch in tissue, depending on its energy. It can penetrate into the living layers of the skin and, thus, is an external as well as an internal hazard to the body.

Let's consider what happens in beta decay. Consider as an example cobalt-60. Cobalt-60 is radioactive and decays by emitting a beta particle, followed by the emission of two gamma rays. Unlike the case with alpha decay, the mass of the nucleus hardly changes since the beta particle's mass is so small. However, the number of protons increases by one (since a neutron has now changed into a proton). The nucleus contains one more positive charge as a result of ejecting the negative beta particle. Cobalt has an atomic number of 27. After it decays, the atomic number of the daughter will be 28. The element whose atomic number is 28 is nickel; therefore, the daughter will be Ni-60. Note that the mass number is still 60 since the sum of the number of neutrons and protons has not changed.

From the changes in mass involved in a particular beta decay, a physicist might suppose that it should be possible to calculate the kinetic energy of the emitted beta particle from Einstein's equation, $E = mc^2$. Lo and behold, however, this is not so. Only the ***maximum*** beta energy can be so calculated; the reality is that the beta energies emitted cover a spectrum of energies. What to do? Clever physicists postulated that, to conserve energy, there must be another "thing" emitted. This thing is the ***neutrino***. This neutrino has no mass but has just the amount of energy given by the difference between the maximum possible beta energy and the actual beta energy. Indeed this is the origin of the neutrino; it always accompanies beta decay. It is difficult to detect, but it has been detected and thus its existence has been verified. The maximum energy of the beta from the decay of Co-60 is 0.32 MeV.

Positrons

As discussed earlier, the positron is a positive beta particle, or positive electron, whichever way you want to think of it. It is identical in mass to the electron but has a single positive charge. It is produced either by radioactive decay or by a process described below called pair production. A positron doesn't stay around long in nature; it very quickly (within microseconds) finds a negative electron to combine with in a process that "annihilates" both particles and produces two gamma rays in their place, traveling in opposite directions. The energy of the gamma rays is equal to the mass of a stationary electron and positron, each of which is 0.51 MeV.

Gamma Rays

A gamma ray is formed from a nucleus that is in an ***excited state***. When the radioactive nucleus returns to a new excited state or to its ***stable***, or ***ground state***, it emits a gamma ray. The excited state has an energy in excess of the ground state, called an ***energy level.*** The energy of the gamma ray is the difference in energy levels between the initial and final states.

Gamma rays can penetrate great distances in both air and tissue. Thus they are a hazard whether they originate outside or inside the body.

Gamma rays, also called gamma ***photons***, are a form of electromagnetic radiation, like light, ultraviolet radiation, infrared radiation, microwaves, and x-rays. Electromagnetic radiation is like a wave, with a wavelength and an oscillating frequency. On the other hand, these forms of radiation exhibit certain characteristics of particles—which appears to be a paradox. The fact that a wave can act as a particle was established in 1900, when the great German physicist Max Planck resolved a baffling discrepancy about thermal radiation by treating the radiation as a particle instead of a wave. This discovery was a breakthrough in the field of quantum physics. A quantum is a particle that contains a discrete amount of energy, and a photon is a quantum of electromagnetic radiation.

Like all electromagnetic waves, gamma rays travel at the speed of light. Gamma rays and x-rays have short wavelengths and high frequencies, as illustrated in figure 2.2. Gamma rays have the highest energies of any of these forms of electromagnetic radiation. The energies of the two gamma rays emitted by Co-60 are 1.33 MeV

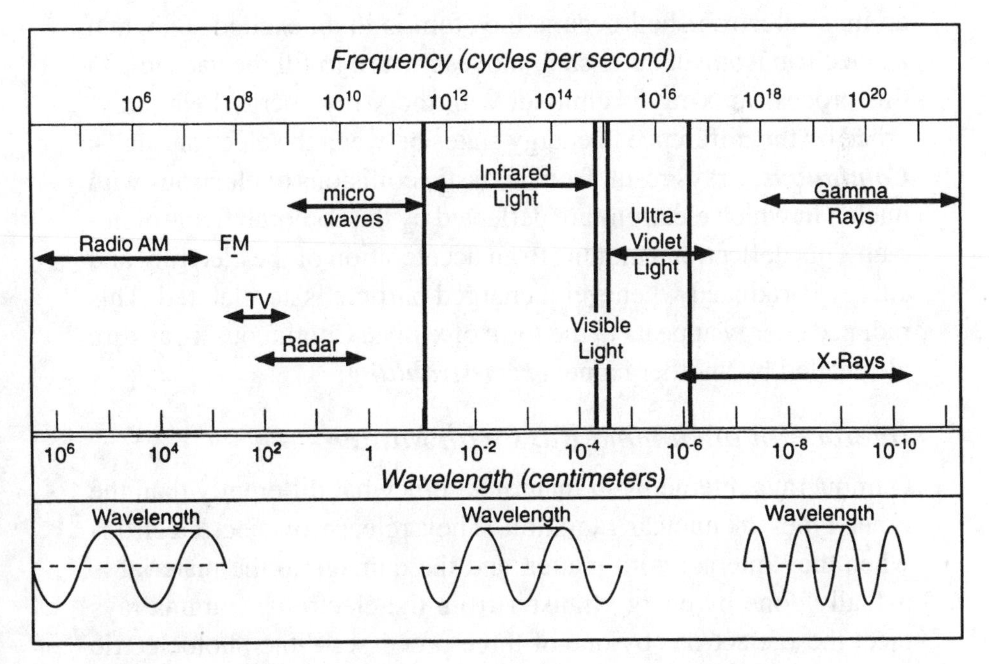

Figure 2.2 The electromagnetic spectrum. COURTESY OF DOE.

and 1.17 MeV. The energy, E, of a gamma ray (and of all photons, for that matter) is related to its frequency, ν, by the relationship: $E = h\nu$, where ν is in hertz (cycles/second), and h is Planck's constant, 4.136×10^{-21} MeV/Hz (or 6.626×10^{-34} J · s). Gamma photons also have momentum despite their lack of mass. While the momentum of normal bodies and particles is the product of mass and velocity, the momentum, p, of a gamma ray is: $p = h/\lambda$, where λ is the wavelength of the gamma ray. The wavelength is the ratio of the speed of light and the frequency, c/ν.

X-Rays

Unlike gamma rays, x-rays do not originate in the nucleus. Yet they do have properties indistinguishable from gamma rays, and they are produced in some types of radioactive decay (electron capture and internal conversion). Like the beta particle and the electron, whether the electromagnetic radiation is called a gamma ray or an x-ray depends on where it comes from. Gamma rays come from the nucleus. X-rays are produced in one of two ways. ***Characteristic*** x-rays come from the electron shells of the atom. Whenever a vacancy in an inner electron shell occurs, the atom is in an excited state, and an electron from an outer shell moves inward to fill the vacancy. In this process an x-ray is emitted, with the x-ray energy being governed by the difference in energy states between the electron shells. ***Continuous*** x-rays result from inelastic collisions of electrons with nuclei, in which electrons are deflected by the coulomb fields of nuclei. This deflection amounts to an acceleration of the electron, and energy is radiated whenever a charged particle is accelerated. This radiated energy appears in the form of x-rays. Continuous x-rays are also called by another name: ***bremsstrahlung***.

Interaction of Gamma Rays with Matter

Gamma rays interact with materials somewhat differently than the other types of nuclear radiation. They release or eject electrons when they interact with matter, and the damage to the material is actually done by energy transfer from the electrons. Gamma rays eject these electrons by one of three processes—the photoelectric effect, Compton scattering, and pair production. These three processes are summarized in figure 2.3.

In the ***photoelectric effect***, first explained by Einstein, the gamma ray is absorbed, transferring all of its energy to an orbital electron that is ejected. In the ***Compton effect***, part of the energy and momentum of the gamma ray is transferred to an orbital electron in an atom, the electron is knocked out of the orbit, the gamma ray proceeds in a new direction at a reduced energy, and the target atom is left as a positively charged ion since it has now lost an electron.

Pair production is strange. In pair production the gamma photon disappears—just vanishes completely—and in its place are two new particles—an electron and a positron. New mass has been created out of the gamma ray's energy! The electron and positron have one negative and one positive charge, respectively, so charge is conserved since the gamma ray had no charge to begin with. All of the energy from the gamma ray cannot be converted into the kinetic

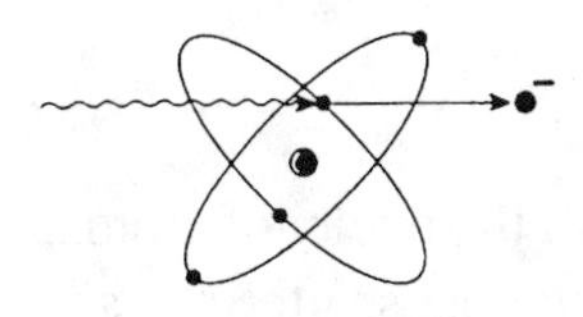

PHOTOELECTRIC EFFECT
- o Gamma ray is completely abosorbed.
- o Electron is ejected with gamma ray's energy minus binding energy.

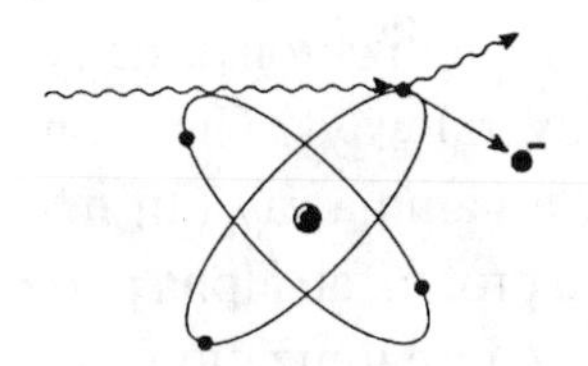

COMPTON EFFECT
- o Gamma ray of lower energy proceeds in new direction.
- o Electron is ejected with the the energy difference.

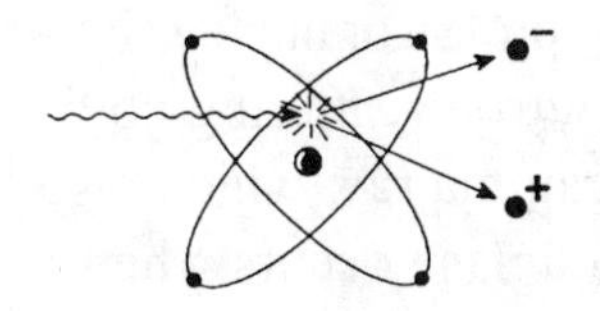

PAIR PRODUCTION
- o Gamma ray is annihilated.
- o Electron and positron are created and share gamma ray's energy minus 1.02 MeV.

Figure 2.3 Interaction processes for gamma rays. COURTESY OF U.S. DEPARTMENT OF ENERGY

energy of the electron and the positron since some of the gamma ray energy has been converted into mass. 1.02 MeV of the gamma ray energy must go into creating the mass of the electron and the positron (from $E = mc^2$): 0.51 MeV for the electron, and 0.51 MeV for the positron. The rest of the gamma ray's energy is shared equally by the two particles as kinetic energy. The positron soon collides with an electron, and both are annihilated to produce two gamma photons. (I find pair production to be one of the most unbelievable processes I know of, but it really happens, all the time! What an exciting confirmation of the equivalence of mass and energy it must have been for the people who discovered it.)

High-energy gamma photons have the highest probability of causing pair production. Low-energy gamma rays are most likely to interact by the photoelectric effect. Intermediate energy gamma rays are most likely to interact by the Compton effect. Very high-energy gamma photons can react in still another way—***photodisintegration***—in which the gamma ray is absorbed by a nucleus that consequently emits a neutron.

Neutrons

Neutrons are neutral particles that collide with the nuclei of atoms. They can undergo a number of reactions, such as scattering, absorption, or, in the case of heavy atoms, fission. Fission is described in some detail in chapter 4. Neutrons do not cause ionization directly in a collision, but they can cause damage in several ways. They can be captured by a material and release either a gamma ray (an n,γ reaction) or one of several particles including a proton, an alpha particle, or even two neutrons; these products can cause ionization and resulting damage. Neutrons can scatter from a hydrogen or heavier nucleus, and the slowing down of the resulting proton or nucleus ionizes the material, causing damage. Some neutron scattering collisions are accompanied by the release of a gamma ray, which also causes ionization. A direct form of neutron damage occurs when a neutron knocks a target atom out of its lattice structure in a scattering collision, thus causing permanent alteration of the structure. Neutrons can penetrate deeply in both air and tissue.

The radioactive decay of the neutron referred to earlier is a curious phenomenon. We all know that a nucleus is composed of so many protons and so many neutrons, and these numbers just stay the same forever in a stable nucleus. It is reasonable to ask, then, how can this be if the neutron is always decaying with its half-life of 12 minutes? The answer is that the coulomb field inside the nucleus is so great that the beta particle that is emitted when a neutron decays is generally immediately attracted to a positively charged proton and unites with it to form a new neutron, thus keeping the net number of neutrons and protons constant. If the electron is not captured by another proton, the electron will indeed be emitted from the nucleus as a beta particle, and the number of protons will increase—which is exactly what happens to a radioactive nucleus that decays by beta decay (though the half-life for beta decay does not parallel that of neutron decay).

Nuclear Reaction Equations

Nuclear reactions are described by equations, much the same way as chemical reactions. The number of protons plus neutrons and the total electric charge are conserved. A couple of examples of nuclear reactions are the following:

(1) neutron absorption: ${}^{1}_{0}\mathrm{n} + {}^{59}_{27}\mathrm{Co} \rightarrow {}^{60}_{27}\mathrm{Co}$
(2) radioactive decay: ${}^{131}_{53}\mathrm{I} \rightarrow {}^{131}_{54}\mathrm{Xe} + \beta^-$
(also frequency written as: ${}^{131}_{53}\mathrm{I} \xrightarrow{\beta^-} {}^{131}_{54}\mathrm{Xe}$)

The neutron is written as in equation 1 because it has zero protons (the subscript) and 1 neutron + zero protons (the superscript). In the decay of I-131, a neutron is replaced by a proton plus a beta particle, so that the total charge has not changed. The number of neutrons + protons (131) has not changed, but the number of protons has changed from 53 to 54.

As discussed earlier in relation to radioactive decay, mass is converted into energy, or vice versa, in a nuclear reaction. As an example, the energy involved in the fission reaction is discussed in detail in chapter 4. Other examples appear in chapter 10 on fusion.

EXPONENTIAL DECAY OF RADIOACTIVE MATERIALS

Radioactive materials decay spontaneously. This means that radioactive nuclei will eject alpha particles, beta particles, and gamma rays at random moments in time. Each time a nucleus emits such radiation, we say that the nucleus has decayed. A particular radioactive material, composed of a very large number of similar nuclei, each decaying randomly, decays as a whole in a very predictable manner. The material decays ***exponentially***.

To understand radioactivity, you've got to know about exponential decay. There is no way out. So let's get started, even if this is completely new to you or if you haven't encountered an exponential since high school.

Exponential decay means the following. If N_0 is the amount of a radioactive nuclide at the initial time, then the amount, N, of the nuclide left after it has decayed for a time, t, is

$$N = N_0 e^{-\lambda t},$$

where $e^{-\lambda t}$ is called an ***exponential***. N can stand for atoms, grams, or any measure of amount. The λ is the ***decay constant*** for the radioactive material. The e stands for a number; it is 2.718 . . . . It comes from the branch of algebra that deals with natural logarithms of numbers—which you do not need to understand in order to understand exponential decay. I do provide some algebra on exponentials and natural logarithms in an Appendix to this chapter for those who are interested. All we need to know here is what the equation looks like when we plot N as a function of time on a graph. This is done in figure 2.4. This graph illustrates exponential decay. N_0 is the amount at time zero so that N/N_0 is 1 at $t = 0$.

Half-Life

An interesting point is shown in figure 2.4 when exactly half of the original material has decayed. The time when half has decayed is called, very logically, the ***half-life***, or $t_{1/2}$. The decay constant is related to the half-life by the relation

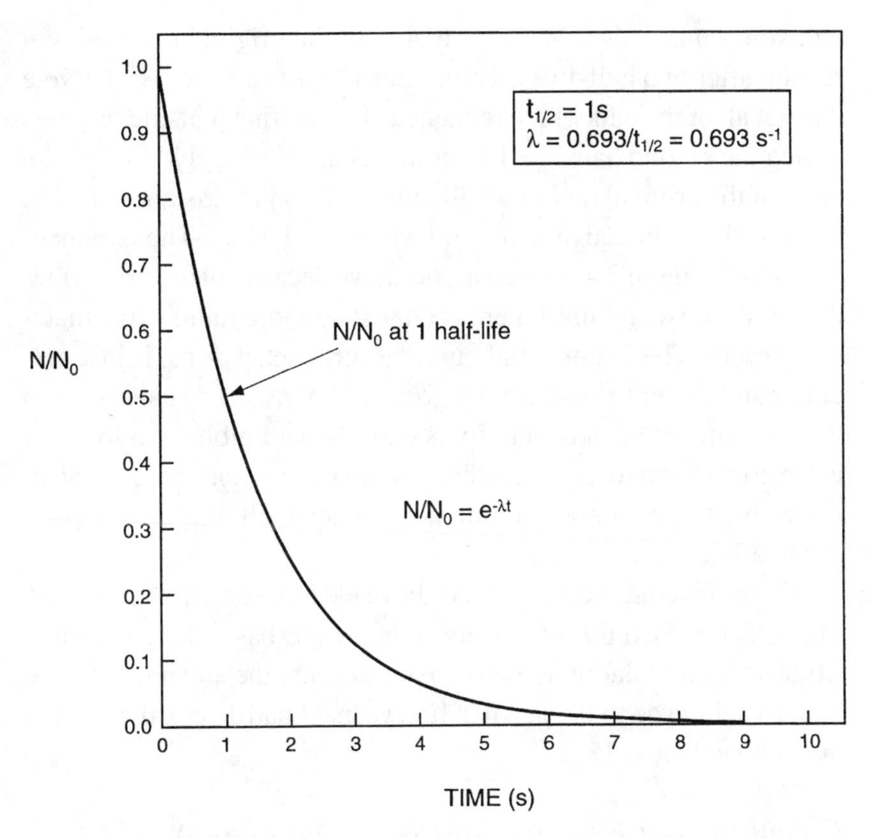

Figure 2.4 Example of exponential radioactive decay (for a half-life of 1 second).

$$\lambda = 0.693/t_{1/2},$$

but you don't really have to know this, or know why (though it is explained in the appendix to this chapter). What is important is the concept of the half-life and the general shape of the decay curve. The time scale for the example in figure 2.4 is seconds. The time scale could just as well have been minutes, hours, days, or years, depending on the half-life of the radioactive material. Half-lives range from billionths of a second to billions of years.

In one half-life, half of the material will have decayed. One half-life later, half of the half that remained after the first half-life will still remain (or ½ × ½), so that ¼ of the original material will remain after two half-lives. Three-fourths will have decayed. (Note that not all of the half which remained after the first half-life decayed during the second half-life!) After three half-lives, only ½ × ½ × ½, or ⅛, of the original material will remain. This process goes on until most of the radioactive material has decayed away. The exponential curve in figure 2.4, as well as the above decay equation, describes this process exactly until there is almost no more radioactive material present. The exponential curve never reaches zero; it just gets closer and closer to zero as time goes on. It may get so close to zero that the amount of radioactivity is entirely negligible—so low that it can do no harm to anyone. Of course, in the end, since we are dealing with a finite number of radioactive nuclei, all of the nuclei will eventually decay.

Consider as an example of the above ideas the decay of cobalt-60. The half-life of Co-60 is 5.3 years. Suppose one has exactly 1 gram of Co-60 at a particular time ($t = 0$). 5.3 years later the amount of Co-60 will have decayed to 0.5 g. After 10.6 years (2 half-lives), the amount of Co-60 will be 0.25 g.

Calculation of Exponential Decay and Activity

Next we ask how much Co-60 is left after some specified time other than a half-life. We must use the above exponential equation to answer this. Let us make the calculation for a time of one year ($t = 1$ y). To use the equation, we must first calculate the decay constant, λ. For Co-60,

$$\lambda = 0.693/t_{1/2} = 0.693/5.3 \text{ y} = 0.131/\text{y}.$$

Substituting this value of λ and $t = 1$ y into the exponential equation gives for the amount of Co-60 after one year

$$\begin{aligned} N &= (1 \text{ g})[e^{-(0.131/\text{y})(1\text{y})}] \\ &= (1)(0.88) \\ &= 0.88 \text{ g}. \end{aligned}$$

Activity, A, the rate at which an amount, N, of radioactive material decays, is governed by the decay constant λ. The equation for activity is

$$A = N\lambda,$$

where N here is specifically the number of atoms. If λ is in units of 1/s, the activity is in disintegrations per second, or becquerels.

As an example of the calculation of activity, consider again the 1 gram of Co-60. The number of atoms in 1 gram of Co-60 is

$$\begin{aligned} N &= (1\ \text{g})(6.022 \times 10^{23}\ \text{atoms/mol}) / (60\ \text{g/mol}) \\ &= 1.0 \times 10^{22}\ \text{atoms,} \end{aligned}$$

where 6.022×10^{23} is Avogadro's number and 60 is the atomic weight of Co-60. In units of 1/s,

$$\lambda = (0.131/\text{y}) \div (60)\ (60)\ (24)\ (365)\ \text{s/y} = 4.2 \times 10^{-9}/\text{s}.$$

Thus, the activity of the Co-60 is

$$\begin{aligned} A &= (1.0 \times 10^{22}\ \text{atoms})\ (4.2 \times 10^{-9}/\text{s}) \\ &= 4.2 \times 10^{13}\ \text{Bq} \\ &= (4.2 \times 10^{13}\ \text{Bq})/(3.7 \times 10^{10}\ \text{Bq/Ci}) \\ &= 1140\ \text{Ci.} \end{aligned}$$

After one year, the activity will be

$$\begin{aligned} A &= A_o[e^{-(0.131/\text{y})(1\text{y})}] \\ &= (4.2 \times 10^{13}\ \text{Bq})\ (0.88) \\ &= 3.7 \times 10^{13}\ \text{Bq} \\ &= 1000\ \text{Ci.} \end{aligned}$$

Some Interesting Radioactive Nuclides

Several interesting radionuclides are listed in table 2.2, along with their half-lives. The naturally radioactive elements listed have long half-lives. This is the reason that U-238 and U-235 are still around in rocks, while Pu-239 is not, and part of the reason why there is so much more U-238 still left than U-235. It is also why potassium-40 is everywhere, including in our bodies, in bananas, and in the earth. Radioactivity is really fascinating when you get to know about it.

Technetium-99m is widely used in nuclear medicine. It decays by isomeric transition, a radioactive process described on page 22. A 140 keV gamma ray is emitted in the process and the nucleus becomes just radioactive Tc-99. This is then followed by the very slow decay of Tc-99 by the emission of a beta particle. A fascinating fact about technetium is that it does not exist in nature. Radioactive technetium is produced from the beta decay of molybdenum-99. Molybdenum-99 is obtained as a fission product from U-235 irradiation in a nuclear reactor.

Sr-90, Cs-137, I-129, and I-131 are fission products (see figure 4.1). These isotopes of strontium and cesium play an important role in governing the time required for the storage of high-level ra-

TABLE 2.2 Several Radionuclides of Special Interest

Radionuclide	*Half-life*
Uranium-238	4.5×10^9 y
Uranium-235	7.1×10^8 y
Plutonium-239	24 000 y*
Hydrogen-3 (tritium)	12 y
Carbon-14	5730 y
Potassium-40	1.3×10^9 y
Cobalt-60	5.3 y
Strontium-90	28 y
Cesium-137	30 y
Iodine-129	1.7×10^7 y
Iodine-131	8.1 d
Technetium-99m	6 h
Technetium-99	2.1×10^5 y

*Note the absence of the comma in 24 000. In the SI system, a space replaces the comma in five-digit numbers. There is no comma or space in four-digit numbers. This convention is used throughout the book. (Also, spaces are used to the right of the decimal place, as was done on page 19.)

dioactive waste from fission. Hundreds of years, i.e., 500 to 1000 years, are required before these fission products decay to safe levels. They are the most radioactive of the long-term fission products, although there are other less active fission products that last much longer, as discussed in chapter 8. If Sr-90 enters the body, it concentrates in the bone and stays there; for this reason it was the nuclide that was of most concern as a possible long-term health hazard from atmospheric testing of nuclear weapons. Iodine-131 is a fission product with an 8.1 day half-life. It is important from the standpoint of weapons testing and also from a possible nuclear reactor accident because iodine can enter the food chain through milk from cows who have eaten grass on which iodine compounds have fallen. Fortunately its half-life is short enough that the danger is diminished after several weeks. Cesium-137 is important in this respect since it can enter the food chain from many foods and for a long time. Like Tc-99, I-129 is a long-life fission product.

Tritium is interesting because it is a principal ingredient in nuclear weapons, i. e., in the hydrogen bomb. Note that its half-life is only 12 years. Therefore, the tritium in a bomb must be replenished periodically because so much of it decays during storage of the weapon. Tritium is made for this purpose in nuclear reactors by the capture of a neutron by deuterium. Tritium is also a fuel in fusion reactors. In a commercial fusion reactor, tritium would be made by neutron capture in lithium, as described in chapter 10.

Carbon-14 is a radioisotope present in all carbon. It is especially useful for dating old organic materials by a process described in chapter 3.

Cobalt-60 is made from the capture of a neutron in nuclear reactors by naturally occurring Co-59. Due to its relatively long half-life, Co-60 is an excellent radioactive source for research and medical use, as in cancer therapy. Co-60 is the main gamma source of radioactive material used in sterilizing medical syringes, cosmetics, feminine hygiene products, and other common materials that most people do not know have been irradiated. (No need to worry, however; these products are not radioactive just because they've been irradiated.) Co-60 and perhaps Cs-137 will also be the sources used when food preservation by radiation becomes widespread (as discussed in chapter 3).

BIOLOGICAL EFFECTS OF RADIATION

So much excessive fear and misinformation exists about the biological effects of radiation that a few words at the outset of this section are in order to provide some perspective on the subject. Exposures to people from the nuclear energy industry and from commercial and medical diagnostic applications of radiation are in the range called ***low-level radiation***. (I will define what is meant by low-level as opposed to high-level radiation shortly.) Based on many statistical comparisons of people who have been exposed to low-level radiation with people who have not, there is no evidence that low-level radiation has any adverse effects on the health of human beings—none whatsoever. The fear of radiation that is so carefully nurtured by some political groups and exploited by the mass media is based on exposure to ***high-level radiation***. The only time that people might be exposed to high-level radiation would be in a nuclear war, or in an accident like Chernobyl (which is virtually impossible in a western-style reactor, as discussed in chapter 6), or in treating cancer with radiation.

Unfortunately, governmental guidelines for radiation protection make people fear radiation even down to extremely low radiation levels. The resulting unwarranted fear has caused serious limitations on the great potential contributions to human quality of life that nuclear energy and radiation have to offer. Some people refuse the medical benefits of radiology and nuclear medicine even though the risks from these low-level exposures are negligible. Fear of risks from low-level waste disposal by the public has delayed the development of low-level waste disposal sites to the extent that the beneficial use of nuclear medicine is being curtailed in many places.

I will begin by describing what happens when tissue is exposed to ionizing radiation. Alpha and beta particles knock electrons from atoms of tissue as they slow down. Gamma rays and x-rays cause ionization by knocking out electrons from atoms; these ***primary*** electrons then knock out other electrons, called ***secondary*** electrons. Electrons are slowed down by interacting with other electrons in the tissue, and the energy lost during these collisions is transferred to the tissue. Neutrons do not cause ionization directly. Instead, neutrons collide with nuclei in tissue, usually hydrogen nuclei (protons). These protons cause ionization, with resulting energy transfer to the tissue. In addition, some neutrons combine with the protons in hydrogen and emit gamma rays. A few neutrons also get absorbed in nitrogen nuclei, emitting protons. Less well established are molecular radiobiological models (not discussed here), which provide details of how these interactions affect human health.

Gamma rays can penetrate deep into the body, even from an external source, and can cause damage to tissue. Beta particles can barely penetrate the skin, but they can cause skin burns; also, beta emission is usually accompanied by gamma emission. Alpha particles cannot penetrate the skin; thus biological damage directly from alpha particles is limited to alpha emitters inside the body. However, alpha particle emission is often accompanied by gamma ray emission so that gamma ray effects can accompany alpha emission, even from external alpha sources.

Radiation Quantities and Units

Radiation quantities and units related to biological effects are complicated—I warn you in advance—but here is a summary of the main points for those of you who want to know.

When x-rays were first discovered, one of their easiest characteristics to measure was the amount of ionization they produced in air. It was natural when physicists defined the unit for x-rays in 1928, to base the unit on the amount of ionization in air and to call that unit the roentgen (R). Physicists would have preferred to have a unit that measured energy deposited in tissue by x-rays, but it was too difficult to measure. One of the main problems with measuring radiation in terms of ionization was that the roentgen did not correspond to the same amount of energy deposited in all tissue or other materials. For example, the energy absorbed in bone from an x-ray is about four times greater than the energy absorbed in soft tissue (which is the reason why bones appear clear on diagnostic x-rays). Eventually physicists learned how to calculate the amount of energy deposited in tissue from measurements of ionization in air. At the present time the roentgen ionization unit is being discouraged, and absorbed dose in tissue is calculated from ionization wherever possible. You still see the roentgen unit in use, and also the rem and the rad. It was convenient for x-rays that, for practical purposes, the numerical values of the roentgen, rad, and rem are the same for radiation striking the body.

The amount of radiation energy absorbed per kilogram in a material is called ***absorbed dose***. The SI unit for absorbed dose is the ***gray*** (Gy). The older unit is the ***rad*** (short for radiation absorbed dose). One gray is equal to 1 joule of energy absorbed per kilogram. The gray and the rad are related: 1 Gy = 100 rad. Dose absorbed throughout a person's entire body is referred to as ***whole-body dose***.

The most common radiations—x-rays, gamma rays, and beta particles—have about the same effect on tissue. However, heavier radiation particles such as neutrons, protons, and alpha particles do considerably more damage on the average per unit of energy deposited. This effect is known as the ***relative biological effectiveness*** (RBE) of the radiation. Thus, heavy particles generally have a higher RBE than gamma rays. To place all types of radiation to tissue on an equivalent basis for radiation protection purposes, a new quantity was needed, called the ***equivalent dose.*** Its unit is the ***sievert*** (Sv). The original unit was the ***rem,*** which stands for rad equivalent man. (I might add in response to women's possible reaction to this unit that the rem was invented in the days before women wanted their equal share of radiation damage!) The sievert is related to the gray by ***quality factors***, Q, with different quality factors assigned to different types of radiation. For example, fast neutrons were given a quality factor of 10 and alpha particles 20. The rem was related to the rad by the same quality factors. Multiplying the absorbed dose in grays by the quality factor for the radiation in question gives the equivalent dose in sieverts.

The sievert and the rem are related: 1 Sv = 100 rem. Often the millisievert (mSv) is used, where 1000 mSv = 1 Sv. Low-level radiation refers to radiation doses in the range of fractions of a mSv to a few hundredths of a Sv (or from millirems to a few rems). A typical chest x-ray is about 0.2 mSv (20 mrem). Natural background radiation exposure to an individual in the United States (excluding radon) averages around 1 mSv (100 mrem) per year. High-level radiation generally refers to levels above several tenths of a Sv (or above tens of rems).

The use of radiation units is constantly being reviewed and revised by organizations of radiation experts. An international body is the International Committee on Radiation Protection (ICRP). In the United States, the National Council on Radiation Protection and Measurements (NCRP) is a nonprofit organization chartered by the United States Congress to make recommendations about radiation protection. In the 1980s, the NCRP established "Scientific Committee 40" to determine better values for the quality factors of different types of radiation. After studying much published data from a variety of experiments, the committee concluded (in NCRP Report Number 104) that it was not possible to assign a quality factor to each type of radiation because biological experiments gave so many different results.

Even though it is not possible to give a specific number to a biological effect for different radiations, it is still desirable to have some way to express the usually greater damage from heavy particles. In NCRP Report Number 116 (1993), the NCRP changed the name of the quality factor to the ***radiation weighting factor***, w_R. In this report they used for w_R values the same as the previously used quality factors even though the scientific validity of these numbers had been questioned by NCRP 104. My intent in reviewing this history is to illustrate the evolving nature of radiation units. It is appropriate to use the latest ICRP and NCRP quantities and units even though the older units are still in common use.

The two basic radiation protection quantities in use are the ***equivalent dose*** and the ***effective dose***, both reported in sieverts. As discussed above, the absorbed dose and the equivalent dose are now related through the radiation weighting factor as follows:

$$\text{Equivalent dose in Sv} = w_R \times \text{Absorbed dose in Gy.}$$

For example, if you have 1 mGy of alpha particle energy deposited in tissue (i.e., the absorbed dose), the biological effect is expected to be 20 times greater than if the radiation had been x or gamma radiation, and the equivalent dose would be 20 mSv. Note a change in both the numerical value and the unit.

Since in many radiation protection situations the radiation to a worker or to a member of the public is not uniform but instead is concentrated in a particular organ of the body, such as the lungs, it is desirable to convert a partial-body absorbed dose in grays to a whole-body effective dose in sieverts. This is done by calculating the *effective dose* using the *equivalent dose* to each organ multiplied by a ***tissue weighting factor***, w_T. For example, the w_T for lungs and bone marrow is 0.12, and the w_T for the thyroid is 0.003. Thus,

$$\text{Effective (whole-body) dose} = w_T \times \text{Equivalent dose in the organ.}$$

An example of the application of these radiation quantities would be the following. Radon, the radioactive noble gas resulting from the alpha decay of radium, is present in the air. If you breathe in radon, you will normally breathe it out before it has a chance to decay since it decays so slowly. However, when radon decays in air, it forms a radon daughter atom that is a solid instead of a gas, and this radon daughter attaches to a dust particle in the air. When you breathe in dust particles, they tend to stick to the moist lining of the lungs and remain there until they decay in several steps, most importantly by alpha emission. The effective dose to the whole body can be calculated in the following way. Suppose an annual absorbed dose to a person's lung from radon progeny is 1 mGy. Since most of this dose is from alpha particles, this absorbed dose is multiplied by w_R of 20 to produce an *equivalent dose* of 20 mSv. Since the radiation is only in the lungs, it is necessary to calculate

TABLE 2.3 NRC Values for Radiation Weighting Factors, w_R

Radiation	*Radiation Weighting Factor*
Gamma rays, x-rays, beta particles	1
Alpha particles	20
Fast neutrons	10
Thermal neutrons	2

the *effective dose* by multiplying the equivalent dose by the tissue weighting factor, w_T, of 0.12 for the lung, producing a whole-body effective dose of 2.4 mSv.

Values for the radiation weighting factor specified by the U.S. Nuclear Regulatory Commission (NRC) for use in the regulation of radiation are given in table 2.3 (as listed in NRC document "10 CFR 20," described on page 60).

Radiation Effects

Acute vs. Chronic. The biological effects of radiation depend strongly on how fast the radiation energy is deposited. A short, or ***acute***, exposure generally does more damage than a ***chronic*** exposure. A chronic exposure generally refers to exposures over months or years to ***low-level radiation***. Cells demonstrate some recuperative capability after exposure to radiation so that a larger dose can be absorbed by a person over a long period of time without harmful effects than can be absorbed all at once. Thus, the biological damage from acute and chronic doses are considered separately.

Somatic vs. Genetic. Radiation has two general effects on people—***somatic*** and ***genetic***. Somatic effects affect the health of a living individual. Genetic effects may appear in offspring in later generations.

Early vs. Late Somatic Effects. There are two types of somatic effects—***early*** and ***late*** (also known as latent or delayed). Early effects are those that appear in a matter of hours or days. Late effects appear after several years, sometimes twenty or thirty years later. The most important late effect is the induction of cancer.

Somatic Effects

No adverse somatic (or genetic) health effects have ever been detected from low-level radiation. Only high levels of radiation far greater than any encountered in the nuclear industry or in diagnostic uses of radiation cause noticeable biological effects.

Early somatic effects from very high doses include radiation sickness and, for extremely high doses, death. Radiation sickness includes nausea and vomiting, fatigue, fever, and blood changes. High levels of radiation can lead to a reduction in the number of white blood cells and destruction of the bone marrow. Other effects include reddening of the skin, shedding of the intestinal lining, and, at extremely high levels, damage to the central nervous system. The main occasion people would be concerned with effects like these would be in a nuclear war.

Another important somatic effect involves birth defects known as ***teratogenic effects***. Among children of atomic bomb survivors, reduction in IQ due to exposures in utero between the 8th and 13th week of gestation for high-level doses [above 0.25 Gy (25 rad)] is well documented.

The lethal acute equivalent dose of radiation, i.e., the amount that results in death, is given by the term ***lethal dose 50/30***. This means that this dose will result in death within 30 days for 50% of a population exposed to this dose. The lethal dose 50/30 for radiation received over a short period of time is about 4.5 Sv (450 rem) to the whole body. However, larger doses are used safely in medical therapy.

Late Somatic Effects and Dose-Response Curves

Cancer is the most important late somatic effect of radiation. Late effects are difficult to measure accurately, though probably more is known quantitatively about late cancer effects from high levels of radiation than from chemical carcinogens. The time for cancer induction from radiation exposure is 5 to 20 years. It is important to re-emphasize, however, that all evidence for radiation-induced cancer has come from high-level exposures and there is no experimental evidence that any cancer is induced by low-level exposure.

Dose-response curves are plots of cancer risk or incidence versus equivalent radiation dose—short-term dose if the exposure is acute, cumulative dose if the exposure is chronic. For exposures to high radiation doses, the history of the Japanese atomic bomb survivors suggests that cancer risk is proportional to the equivalent dose, so that the dose response is ***linear***. At low doses, however, there is considerable uncertainty about the shape of the dose-response curve. This uncertainty is important because dose-response curves for high doses are used to establish standards for ***radiation protection*** for low-level radiation.

The four principal dose-response curves are (1) ***linear, no-threshold***, (2) ***linear-quadratic***, (3) ***threshold***, and (4) ***hormesis***. These are illustrated schematically in figure 2.5. The solid circles are representative of actual data obtained at high doses. High-level doses are tens to hundreds of times higher than low-level exposures. The low-level radiation region in the lower left corner is a very small segment of this figure. The number of cancer deaths among people in the high-dose region of figure 2.5 is large enough that scientists can determine statistically that some of the cancers were really caused by radiation. The curves at the low level must be postulated from high-level radiation data and dose-response theories because no cancer cases have ever been identified from low-level radiation. It is difficult (or in some cases impossible) to obtain statistically significant results on the number of cancers that might result from exposure to low-level radiation because this number is either zero or is insignificant compared with the total number of cancers from other causes.

In the Rocky Mountain states, where non-radon radiation levels are about double those of the rest of the country, the cancer rates are 15% lower than the U.S. average. Examples in other parts of the world of this lack of correlation between high background radiation levels and cancer are discussed later in the section on background radiation.

The main source of data for human cancer dose-response curves is the histories of survivors of the atomic bombs dropped on Japan in the Second World War. Most of this radiation was from gamma

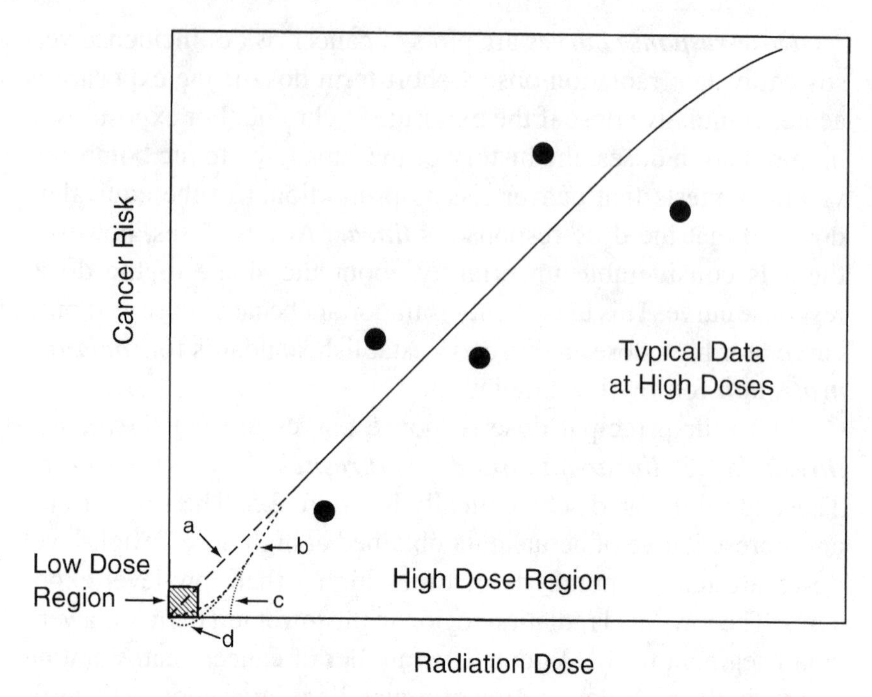

Figure 2.5 Schematic dose-response curves: (a) linear, no-threshold (b) linear-quadratic, (c) threshold, (d) hormesis.

rays, though some was from neutrons. There is no experimental evidence from these or any other histories that cancer deaths are caused by acute radiation doses below about 0.2 Sv (20 rem). The cancer rate for atomic bomb survivors who received 0.005 to 0.05 Sv was actually lower than for those who received no radiation. Hence, if that data point were plotted on figure 2.5, it would fall below zero, that is, on the hormesis curve!

The linear, no-threshold dose-response curve in figure 2.5 is simply a linear extrapolation of cancer incidence from high dose down to low dose where no data exist. It is the most conservative of the four curves on figure 2.5, meaning that it demonstrates the highest cancer risk for low-level radiation. It assumes that any amount of radiation is dangerous, no matter how small, and it ignores the body's repair mechanisms for exposure to radiation. There is no theoretical basis in humans for the linear, no-threshold dose-response curve, and there is no scientific evidence to support it. Such a large extrapolation to low doses without any biological model to support it is tenuous indeed. Nevertheless, the linear dose-response curve is so widely used in establishing radiation protection standards that it has a name. It is called the ***linear, no-threshold hypothesis*** or, sometimes, simply the linear hypothesis. Many radiation scientists are convinced that the linear, no-threshold hypothesis for low-level radiation is invalid and should be abandoned.

In the case of the radium dial painters (discussed in the history section of chapter 4) there is overwhelming evidence that a ***threshold*** exists below which no additional cancers were caused by the alpha-emitting radium they ingested. The exhaustive research on cancer incidence in radium dial painters was performed over a forty-year period by Professor Robley Evans and his associates at the Massachusetts Institute of Technology (MIT). His research involved the incidence of radiogenic cancer tumors, meaning tumors caused by the radium deposited in bones. Chemically, radium behaves like calcium, and an appreciable fraction of ingested radium is deposited in the skeleton. In 1974, in the journal *Health Physics*, Evans showed a dose-response curve, which is reproduced in figure 2.6. His results clearly show a threshold at about 10 Gy (1000 rad) to the bone, a very large radiation level indeed. Evans followed the histories of approximately 400 painters who ingested radium. None of the more than 300 who received cumulative doses below 10 Gy (1000 rad) developed radiogenic tumors; 19 of the 67 who received more than 10 Gy developed radiogenic tumors. Histories of surviving dial painters have been followed at the Argonne National Laboratory during the twenty years since Evans completed his research, and the threshold still exists. The doses received by the dial painters were delivered over decades; hence, the exposures were chronic. It appears that time was available for the body to repair this radiation damage at doses below 10 Gy. (For reference, using the appropriate values of w_R (20 for alpha) and w_T (0.12 for bone), this 10 Gy (1000 rad) threshold for alpha radiation to the bone corresponds roughly to an effective dose to the whole body of 24 Sv (2400 rem).)

The fourth dose-response curve, hormesis, is receiving increasing attention. This theory says that radiation at low levels is

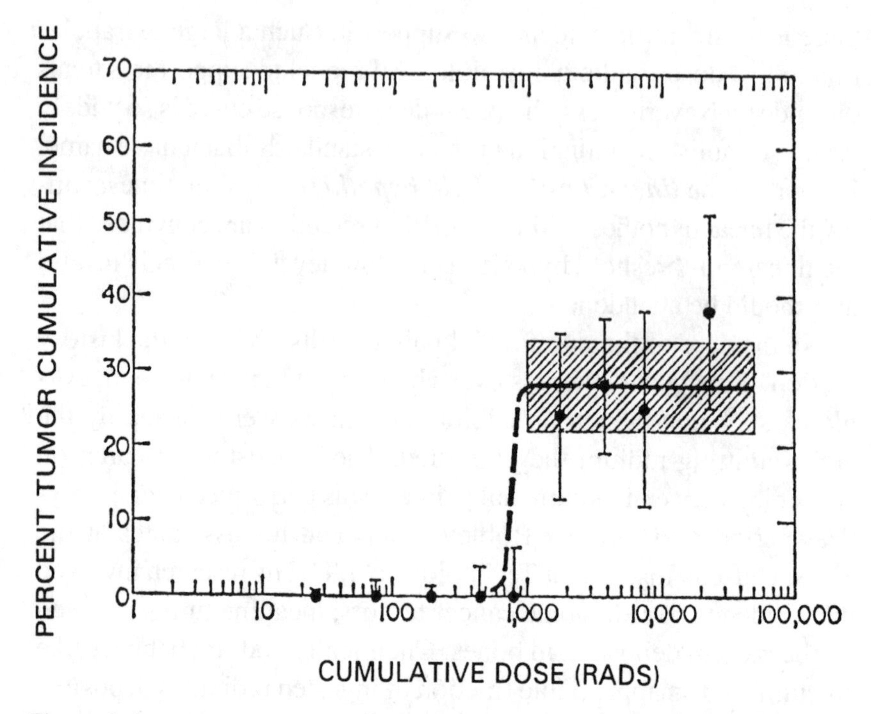

Figure 2.6 A threshold dose-response curve, showing radiation-induced bone tumor incidence for radium dial painters. The shaded region corresponds to the average occurrence, 28 ± 6% between 1000 and 50 000 rads (10 and 500 Gy). Note that the dose scale is "logarithmic." This means that each increase by a factor of ten in the cumulative dose is represented by the same length on the graph. A normal ("linear") scale was used in figure 2.5. *Health physics,* November 1974. Reprinted with permission.

actually beneficial to people—as contradictory as this sounds. Many things in small doses are beneficial, for example, minerals and vitamins. There is much evidence to support the idea that this also applies to radiation. For example, I have already referred to the 30 000 atomic bomb survivors who received between 5 and 50 mSv (0.5 to 5 rems). The number of cancer deaths in this group is statistically significantly lower than the number expected (lower by 6%). The number expected is obtained by comparison with cancer deaths among people in a control group from the same area who received no radiation dose. Moreover, the exposed survivors are outliving the non-exposed survivors. Similar health benefits for various groups of occupational workers exposed to low-level radiation have been found. There is a substantial body of evidence indicating that low-level radiation stimulates biological defense mechanisms against further damage by low-level radiation, but the evidence is not yet sufficiently well established to influence radiation standards.

Professor Bernard Cohen has recently shown that the effects of radon in houses exhibits hormesis. This contradicts the linear, no-threshold hypothesis currently used by the U.S. Environmental Protection Agency (EPA) for predicting thousands of lung cancer deaths from radon in homes. Cohen showed that lung cancer mortality rates are actually lower in counties with higher radon levels, even after accounting for the effects of smoking. Results for lung cancer mortality were plotted by Cohen for 1601 counties in the United States as a function of average radon level in homes. This plot for females is shown in figure 2.7. The lower scale is r/r_o, where r_o is a widely used radon concentration unit of 37 Bq/m^3 (1.0 picocurie/litre). (The EPA has set 4 pCi/L as a level of concern.) The solid line was drawn through the data by Dr. Cohen. The dashed line is the linear, no-threshold hypothesis used by EPA for setting radiation protection guidelines. Since the solid line (the actual data) slopes downward, it appears that radon radiation at low levels has a beneficial effect, while the linear, no-threshold hypothesis appears to be invalid for low-level radon exposure. For males the difference between the data and the linear, no-threshold hypothesis is even more pronounced.

The data in figure 2.7 appear to flatten out between 3 and 7 r/r_o. It is known from high-level exposures to uranium miners in the early days of the uranium mining industry that high-level exposure to radon does lead to excess lung cancer. Therefore, if figure 2.7 were extended to exposures much greater than r/r_o of 7, the dose-response curve would rise again in a shape that radiologists refer to as a "J shape."

In order to ensure the health and safety of the public, the NCRP recommends guidelines for government regulation of radiation in a conservative manner. This means that a pessimistic ***upper limit*** to

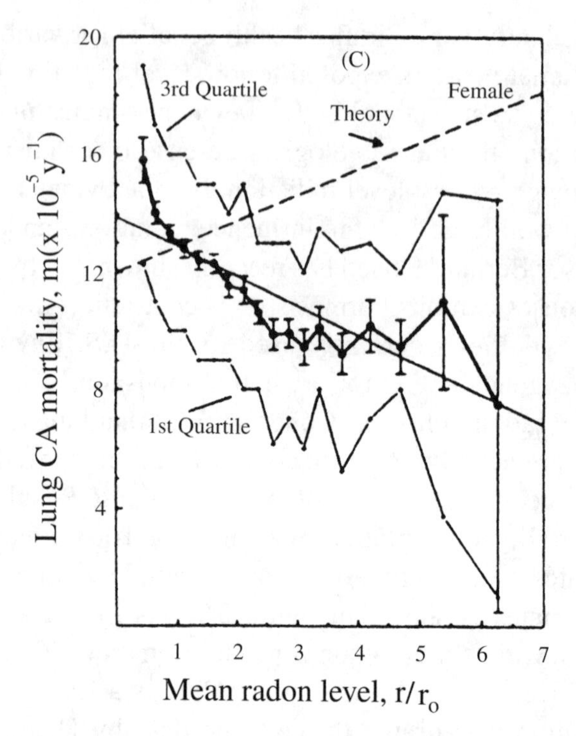

Figure 2.7 Lung cancer mortality among females versus average radon levels for 1601 U.S. counties. *HEALTH PHYSICS,* FEBRUARY 1995. REPRINTED WITH PERMISSION.

risk is established. This upper limit guides government regulations for the protection of the public. To obtain an "upper limit" on radiation damage for radiation protection purposes, the NCRP has adopted the linear, no-threshold hypothesis.

Statistical data collected over the last few decades provides increasing evidence that the linear, no-threshold hypothesis in invalid. One might think at first glance that this ultra-cautious approach can do no harm—if in doubt, best to err on the side of caution. However, this is not so. If the linear, no-threshold hypothesis is indeed invalid, its use as a radiation protection standard is extremely harmful. Its use leads to huge expenditures of both private and tax dollars to reduce radiation levels. (Some examples of excessive costs involved in radiation protection are given in chapters 8 and 9.) If the hypothesis is invalid, these resources could be better applied to real health and safety issues. The hypothesis that radiation is dangerous down to any level feeds the public's fear of radiation to the extent that many benefits of radiation, such as nuclear medicine and nuclear energy, are dramatically curtailed.

Numerically, the linear, no-threshold hypothesis places the probability of a radiation-induced cancer death of 1 (i.e., certain death) at a chronic whole-body dose of about 20 Sv (2000 rem) and an acute dose of about 10 Sv (1000 rem). This includes all types of cancer, such as leukemia and cancer of various organs. These numbers are recommended by the NCRP in their report ***Risk Estimates for Radiation Protection***, NCRP Report No. 115, published in 1993. The NCRP emphasizes that these recommendations are ***for radiation protection purposes***. It is a misuse of the linear, no-threshold hypothesis to use it to predict numbers of cancer deaths from low-level exposures.

To understand how the linear, no-threshold hypothesis is applied, let's talk about chronic doses for now. The linear, no-threshold hypothesis is equivalent to saying that for each sievert of chronic radiation absorbed by an individual, there is 1 chance in 20 that this person will eventually die of cancer as a result of this radiation exposure. (Recall that 1 Sv is high-level radiation.) The linear, no-threshold hypothesis says much more, however, when we introduce the concept of a ***person sievert***. The person sievert received by a group of people is equal to the total dose received by the group; i.e., the sum of the doses received by each person, or, equivalently, the product of the average dose per person and the number of persons. Suppose that instead of one person receiving 1 Sv, 10 persons each receive 0.1 Sv, or 100 persons receive 0.01 Sv. The cumulative dose to the population in the three cases is the same; namely, 1 person sievert. Due to the assumed additive risk of radiation inherent in the linear hypothesis, no matter how small the exposure, this hypothesis says that the chance of a cancer per person sievert is the same (1 in 20), regardless of the size of the group or the average dose received. For the case of 10 persons each receiving 0.1 Sv, the chance that one of the ten will contract cancer is still 1 in 20 according to the linear hypothesis, but now the chance of a specific individual contracting cancer is 1 in 200. For the 100 persons each receiving 0.01 rem, the chance of cancer for a specific person is one in 2000 according to the linear hypothesis. You can see that the linear, no-threshold hypothesis says that the probability that a single cancer will occur is the same whether that sievert was absorbed by a single person or distributed in small doses over many persons. Again, it is worth emphasizing that if all the person sieverts are the result of exposures to low-level radiation only, there is no scientific evidence that this exposure would result in any adverse health effects, in apparent contradiction to the application of the linear, no-threshold hypothesis to low-level exposures.

Another way to understand the linear, no-threshold hypothesis is to use it in absurd situations to which we all know that it does not apply at all! If one takes many aspirin tablets at one time, the result will be fatal. Let us suppose, for illustration, that the fatal number is 100. The linear hypothesis says that if 100 people each take 1 aspirin, one will die from the aspirin. Another absurd example: if one falls a certain distance, it will surely be fatal. Let us suppose that the surely fatal distance is 100 feet. The linear hypothesis says that if 100 people each fall one foot, one will die from the fall. Radiation is different from these absurd examples because any amount of radiation has the potential to damage an individual cell. Since repair mechanisms for radiation and how cell damage might lead to cancer are not fully understood, the radiation protection committees take the cautious, conservative approach of the linear, no-threshold hypothesis for radiation protection standards.

The linear, no-threshold hypothesis is sometimes misused to estimate the maximum number of cancer fatalities from a nuclear accident. You may have heard predictions that thousands of fatalities might eventually result from the Chernobyl nuclear accident. These predictions are based on estimates of person sieverts of radiation received as a result of the accident, together with the use of the linear, no-threshold hypothesis. There were 600 000 clean-up workers at Chernobyl, many of whom received doses in the range of 0.2 Sv. These doses are high enough that some cancers may result among this group from their radiation exposure. Doses to individuals in the surrounding populations, however, were low enough that it is unclear whether any cancers will be caused in this group by their radiation exposure from Chernobyl. Even the number of deaths of members of the public predicted by misuse of the linear, no-threshold hypothesis may be impossible to detect since this number would be so small compared with the number of natural cancer deaths. So far the only reported cancer effect that appears to be real is some increase in thyroid cancer in children from iodine-131. Fortunately, thyroid cancer is one of the most curable types of cancer. There has been no significant increase in leukemia deaths, even though many were seen in the atomic bomb victims.

I recommend that you be skeptical when you hear media reports that thousands have already died from Chernobyl. Comparison with the results of the atomic bomb survivors from Hiroshima and Nagasaki quickly illustrates how absurd recent claims of high death rates are. There have been about 400 excess cancer deaths among the 76 000 atomic bomb survivors whose health histories were followed during the 50 years since the war, and most of these occurred after ten years following the radiation exposure. Recall that "excess deaths" is the term for the increase in the number of cancer deaths over the expected number. These excess cancer deaths account for about 1% of the total deaths so far among the atomic bomb survivors. Physics Professor Richard Wilson of Harvard University, one of the most knowledgeable scientists involved in assessing health effects from Chernobyl, points out that the cumulative radiation received by members of the public living within 30 kilometres of Chernobyl (the area evacuated after the accident) was about the same as that received by the Japanese atomic bomb survivors, approximately 10 000 person sieverts. Hence, members of the public who lived near Chernobyl should not be expected to develop more excess cancers than the atomic bomb survivors. People living 30 kilometres away from Chernobyl received only very low levels of radiation. The cumulative dose to the Chernobyl clean-up workers was about ten times the amount that the surrounding public received, and many of the doses received were high-level rather than low-level doses.

Therefore, radiation-induced cancer incidence among this group may be greater than for the public. Most of these cancers will appear more than ten years after the accident.

Genetic Effects

Genetic effects of radiation in humans tend to be blown all out of proportion. Radiation-induced genetic effects have been shown to exist in fruit flies and mice, but ***they have never been seen in humans***. Even after following 90 000 children and grandchildren of the Hiroshima and Nagasaki survivors for a half-century, ***there has been no statistically significant increase in genetically related disease in the offspring of atomic bomb survivors***. This comes as a great surprise to almost everyone I tell this to. The media have conditioned the public to fear the genetic effects of radiation. Most quantitative estimates of the genetic effects of radiation in humans come from extrapolations from studies on mice—millions of them. What examination of the genetic results from the atomic bomb survivors does tell us clearly is that humans are more resistant to genetic defects from radiation than would be expected from mice experiments.

Two types of genetic effects are chromosome abnormalities and gene (point) mutations. Each normal person has 23 chromosomes. There are thousands of genes in each chromosome. Genetic effects can come from defects in either.

The concept most widely used to assess genetic effects of radiation is the ***doubling dose***. This is the radiation dose needed to double the natural rate of genetic disorders in humans. This doubling does not need to occur in the first generation; the theory is that the doubling would take place over many generations, for several centuries, until a new plateau is reached that is twice as high as the present one. The doubling dose recommended by the most recent (1990) National Research Council's Committee on Biological Effects of Ionizing Radiation (BEIR V) is 1 Sv (100 rems) of low LET radiation. The Committee acknowledges a large uncertainty in this value and considers it to be close to a lower limit since it is based on mice experiments. The actual doubling dose may be several sieverts. It is also generally accepted from mice and fruit-fly studies that genetic effects vary linearly with dose.

There are many types of genetic defects, some important and many minor. As indicated by the concept of doubling dose, a comparison with natural genetic defects is often used to gain some perspective on the role of radiation. I have seen varying estimates for the rate of natural genetic defects from all causes, generally around 4% of live births, or 40 000 per million births. Only a small percentage of these result from radiation to which the average American is ordinarily exposed; an estimate consistent with the BEIR V report is that 3% of genetic defects are caused by radiation. The BEIR V report estimates that for a population uniformly exposed to 1 mSv of low LET radiation, there would be 10 to 20 genetic defects per million live births per generation. Bernard Cohen, whose book I referred to in the preface, estimates that a total of 1 genetic defect in all future generations (not 1 in each generation, but 1 total) is produced for each 110 person sieverts of equivalent dose absorbed by the population. The National Council on Radiation Protection and Measurements (NCRP) recommended in 1993 that an assumption of 1 "severe hereditary effect" for each 100 person sieverts of equivalent dose be used for purposes of radiation protection. A large nuclear energy industry will result in less than 0.01 mSv of radiation per person per year. Clearly, any increase in genetic defects from such a minute addition to the radiation that we already receive (as described in the section below on natural background radiation) is negligible.

PROTECTION AGAINST IONIZING RADIATION

Three fundamental techniques are used to protect people from ionizing radiation. These are the use of (1) ***time***, (2) ***distance***, and (3) ***shielding***.

Using ***time*** for radiation protection means simply limiting the amount of time a person is exposed to a source of radiation. A source generates particles at some rate, in particles per unit time; thus, the dose absorbed by a person near a source is proportional to the amount of time the person is near the source.

Distance can be used because the dose rate from a source decreases with the distance from a source; thus, the greater the distance from the source, the lower the dose. For example, the dose rate from a "point source," i.e., radiation coming from a small source, follows an ***inverse square law***, meaning that the dose rate decreases with distance, r, from the source as $1/r^2$.

Shielding is material placed around or near a radiation source to protect people from the source. Placing shielding between a source and a person will reduce the dose to the person. High-density materials like lead are the best shields for gamma rays and x-rays. Hydrogenous materials, like water, combined with good neutron absorbers, like boron, provide good shielding from neutrons. Concrete is an inexpensive shield widely used for both gamma rays and neutrons. You may have experienced the use of shielding during an x-ray when a dentist or x-ray technician placed a lead apron over part of you. You may have also noticed that the technician moves away, sometimes to a position behind a wall, when the x-ray is taken.

At the beginning of this chapter and of chapter 1, I pointed out that people who work with nuclear energy do not fear radiation. The reason is that they understand these three techniques for protection against radiation, and they make use of them in various combinations all the time.

NATURAL BACKGROUND RADIATION

Radiation is everywhere in our environment. The average American is subjected to about 3.6 mSv (360 mrem) of radiation per year. A breakdown of the sources of background radiation is given in table 2.4, which are values obtained from the BEIR V report.

The largest source of natural background radiation is radon. Most of the significant radiation actually comes from the solid daughter products, or progeny, of radon—various isotopes of polonium, lead, and bismuth. When you hear about the "effects of radon," this includes the progeny. Radon is an inert gas. There are isotopes of radon in both the uranium and the thorium decay chains. Since uranium and thorium are everywhere in the ground, radon gas comes up from the ground into buildings. On average, Americans receive about 2 mSv (200 mrem) per year from radon and its progeny. However, radon

TABLE 2.4 Average Annual Effective Dose to a Member of the U.S. Population

Source	Annual Effective Dose (mSv)	(mrem)	%
Natural			
Radon	2.0	200	55
Cosmic	0.27	27	8
Terrestrial	0.28	28	8
Internal (non radon)	0.39	39	11
Total natural	3.0	300	82
Artificial			
Medical			
X-ray diagnostic	0.39	39	11
Nuclear medicine	0.14	14	4
Consumer products	0.10	10	3
Occupational	< 0.01	< 1	< 0.3
Nuclear fuel cycle	< 0.01	< 1	< 0.03
Fallout from weapons testing	< 0.01	< 1	< 0.03
Miscellaneous	< 0.01	< 1	< 0.1
Total artificial	0.63	63	18
Total natural & artificial	3.6	360	100

concentrations in homes and workplaces vary greatly so many of us receive relatively little radiation from radon while others may receive several times the national average. The dose from radon and its progeny is confined to the lungs. The 2 mSv is an effective whole body dose, meaning that the health risk is estimated to be the same as if 2 mSv were absorbed throughout the body. The effects of radon are estimated mainly from early uranium miners who worked in high radon concentrations, but the validity of extrapolating these effects to low radon dose-rates in homes is highly uncertain, as described in the section on biological effects of radiation.

We are all subjected to about 1 mSv (100 mrem) of non-radon natural radiation each year. In addition, a person in the United States receives an average of about 0.5 mSv (50 mrem) per year from diagnostic and therapeutic medical radiation. Consumer products add another 0.1 mSv (10 mrem).

Natural and artificial radiation exposure to the public, other than radon and medical radiation, comes from several sources. Cosmic rays from outer space (which are, for the most part, protons) account for about 0.3 mSv (30 mrem) per year, and several times that if you live at high altitudes, like Denver. Gamma radiation from uranium, thorium, and their daughters in rocks and soil and building materials adds another 0.3 mSv. A person living in a wooden house generally receives half as much radiation from the building as one who lives in a brick or concrete house. Potassium-40 in our bodies adds about 0.4 mSv. (The dose from carbon-14 in our bodies is negligible.) Fallout from nuclear weapons testing used to add a little, but only cesium-137 and strontium-90 remain, and now the average annual dose from these radionuclides is less than 0.01 mSv (1 mrem). Each time a person takes a cross-country airplane trip, he/she is exposed to 0.02 or 0.03 mSv (2 or 3 mrem) of cosmic radiation. Color TV adds a little, approximately 0.01 mSv per year average. Smoke detectors give an average of about 0.02 mSv/year. A person working in the United States Capitol Building receives about 0.2 mSv/year from the walls; someone working in Grand Central Station in New York receives 1.2 mSv/year. Cooking with natural gas leads to an average exposure of 0.02 mSv/year from the tritium in natural gas. For the average citizen, the radiation exposure added as a result of electricity generation by nuclear energy is completely negligible, being way below 0.01 mSv/year.

If one tried to use the linear, no-threshold hypothesis to estimate the number of cancer deaths in the United States from the average background dose, one would arrive at 40 000 fatal cancers per year. (This is 0.360 rem times 230 million Americans divided by 2000 rems/cancer death.) The number of new cancers per year in the United States is of the order of one million. 40 000 is only 4% of the total, and the variation in cancer rates between different sections of the country are much greater than 4%. Furthermore, in the Rocky Mountain states, where the nonradon radiation levels are about double those of the rest of the country, the cancer rates are 15% lower than the U.S. average.

In some parts of the world, the nonradon natural radiation levels are much higher than the average value in the United States of 1.0 mSv/year (100 mrem/year). In a village in Brazil (Guarapari), the average nonradon average absorbed dose rate is 6 mSv/year (600 mrem/year). Along the Kerala coast of India, the average nonradon background is 4 mSv/year (400 mrem/year). In parts of Guangdong Province in southern China, the background radiation, including radon, is nearly three times as high as in neighboring regions of China. Most of these high dose-rates result from high concentrations of thorium and uranium in monazite sand in the regions. Despite efforts to find correlations with radiation levels in these regions, in none of these cases has it been possible to show that cancer rates are higher than in adjoining regions with lower radiation levels. Such results highlight the extremely conservative nature of the linear, no-threshold hypothesis.

By law, a nuclear power plant must be designed so that the maximum amount of radiation dose that can be absorbed by a person living next to it during normal operation is 0.05 mSv (5 mrem) per year. This can be compared with the average annual per capita dose of 3.6 mSv (360 mrem) in the United States to see how small an amount a nuclear plant adds to the environment during normal operation. Despite this minute addition, nuclear critics frequently publish reports of large increases in radiation-related problems around nuclear plants, which become newspaper headlines and television stories. Invariably, follow-up studies by radiation scientists are able to show the fallacies of the statistics and thus discount the alarming results, which are clearly impossible due to the small releases from the plants. Of course, these later findings either make no news or are relegated to the back pages of the newspapers.

One large study, however, finally made front page news in 1990. Since alarmists continued to complain about damage to the public from normal operation of nuclear facilities, the National Cancer Institute (NCI) was mandated by Congress to study health effects around nuclear power stations and other nuclear facilities operating in the United States. The two-year NCI study, "Cancer in Populations Living near Nuclear Facilities," was published in 1990. The NCI found no evidence of any excess occurrence of cancer among people living in counties that have, or are adjacent to, nuclear facilities in the United States. NCI studied 52 commercial nuclear power stations that began operation before 1982 and 10 other facilities involved in the nuclear fuel cycle, nine of which were defense plants. ***Study counties*** where nuclear facilities were located were compared with ***control counties*** whose populations were similar in income, education, and other socioeconomic factors. The investigation included 107 study counties with or near nuclear facilities in 34 states and 292 control counties. A total of 2.7 million cancers were considered in the comparisons. Sixteen types of cancer were studied, with special emphasis on leukemia. Most of the results compared cancer fatalities. However, in counties where records were available, cancer incidence was also studied. Neither cancer mortality not cancer incidence was higher in the study counties than in the counties without nuclear facilities.

REGULATION AND "ALARA"

Allowable radiation dose limits are specified by the U.S. Nuclear Regulatory Commission (NRC). These limits are based on recommendations by international bodies of radiation experts, such as the International Commission on Radiation Protection (ICRP).

The allowable limits for ***occupational workers*** in nuclear fields are given in table 2.5. The allowable yearly contribution from an NRC licensed operation to the whole-body radiation dose to an individual member of the ***general public*** is 0.1 rem. These limits are given in the NRC document generally referred to as "10 CFR 20," which stands for Chapter 10, Code of Federal Regulations, Part 20. Its title is "Standards for Protection against Radiation." This document also sets limits for every radionuclide on radioactivity concentrations that can be released without restriction to the general water and air environment.

TABLE 2.5 Allowable Radiation Dose Limits* for Occupational Workers in the Nuclear Field

Region of the Body	*Dose Limit (rem/year)*
Whole body	5
Individual body organs other than the lens of the eye	50
Lens of the eye	15
Skin and extremities**	50
Embryo/fetus	0.5 rem during the entire pregnancy

*For parts of the body, the allowable limit is based on the whole-body effective dose in which the actual dose to the organ or tissue is modified to give the same effect as if the dose had been absorbed by the whole body. Special rules apply to summing the dose from multiple organ exposures.
**Extremities include hands, forearms, elbows, feet, ankles, shins, and knees.

A guiding principle for the use of radiation is the term ***ALARA***, which stands for ***as low as is reasonably achievable***. As the name implies, the NRC requires that all who handle radioactive material make a conscious effort to keep personnel exposures to as low a level as is reasonably achievable, a concept that is taken very seriously by people working with radiation. As a consequence, workers rarely approach the allowable radiation limits given in table 2.5.

APPENDIX TO CHAPTER 2

Exponential Equations

We have learned that radioactive materials decay according to the relationship

$$N = N_o e^{-\lambda t}. \tag{A1}$$

where e is the base of the natural logarithm discussed above and $e^{-\lambda t}$ is an exponential. We also learned that

$$\lambda = 0.693/t_{1/2}. \tag{A2}$$

For those interested in the mathematics, I will show how we obtain equation A2. I will also show a way to plot radioactive decay different from, and often more convenient than, the method used in figure 2.4.

We begin with a discussion of logarithms. The symbol ln(x) stands for the natural logarithm of x, based on the number e, where (to four digits) e = 2.718. The natural logarithm and the base e follow several basic algebraic laws, similar to the common logarithm log(x), which is based on 10. Natural logs are easier to handle than common logs, however, since one does not have to worry about mantissas and the lot. Natural logs also appear in equations describing natural events, such as the decay of radioactive materials.

Exponential functions and natural logarithms have the following properties (I'll not begin to tell you why; just believe me—it's so)

$$e^{-a} = 1/e^{a}. \tag{A3}$$

$$\ln(e^{a}) = a. \tag{A4}$$

$$\ln(1/a) = -\ln(a). \tag{A5}$$

Let's re-examine the decay constant, λ. The decay constant has the units of inverse time (1/s). At time $t = 1/\lambda$, equation A1 gives,

$$N/N_o = e^{-1} \quad \text{(A6)}$$
$$= 1/e \text{ [from Eq(A3)]}$$

This means that at a time equal to the inverse of the decay constant, the amount of radioactive material remaining is 1/e, or 1/2.718, or 0.37 of the original amount present.

More interesting is the half-life, $t_{½}$, and its relation to λ. We ask the question, at what time is $N/N_o = ½$? This time, of course, is the half-life. Setting N/N_o equal to ½ and setting the time to $t_{½}$ in equation A1 gives

$$N/N_o = ½ = e^{-\lambda t_{½}} \quad \text{(A7)}$$

We use equation A4 to solve for $t_{1/2}$ after taking the logarithm of both sides:

$$\ln(½) = \ln(e^{-\lambda t_{½}}) \quad \text{(A8)}$$
$$= -\lambda t_{½}.$$

Next we make use of equation A5, which says that

$$\ln(½) = -\ln(2). \quad \text{(A9)}$$

Substituting (A9) into (A8) gives

$$\lambda t_{½} = \ln(2) \quad \text{(A10)}$$

The value of ln(2) is 0.693. Therefore, finally, the decay constant and half-life are related by

$$\lambda = 0.693/t_{½}. \quad \text{(A11)}$$

The shape of a typical radioactive decay curve was shown in figure 2.4. A more convenient way to plot radioactive decay is to plot it on a ***semi-log*** plot, using a special type of graph paper. A plot of equation A1 on semi-log paper results in a straight line. One is actually plotting $\ln(N/N_o)$ as a function of time instead of just N/N_o versus time, as was done in figure 2.4, but the use of semi-log paper makes it so you don't have to calculate the logarithm. As an example, figure 2.8 is a semi-log plot of the decay curve in figure 2.4.

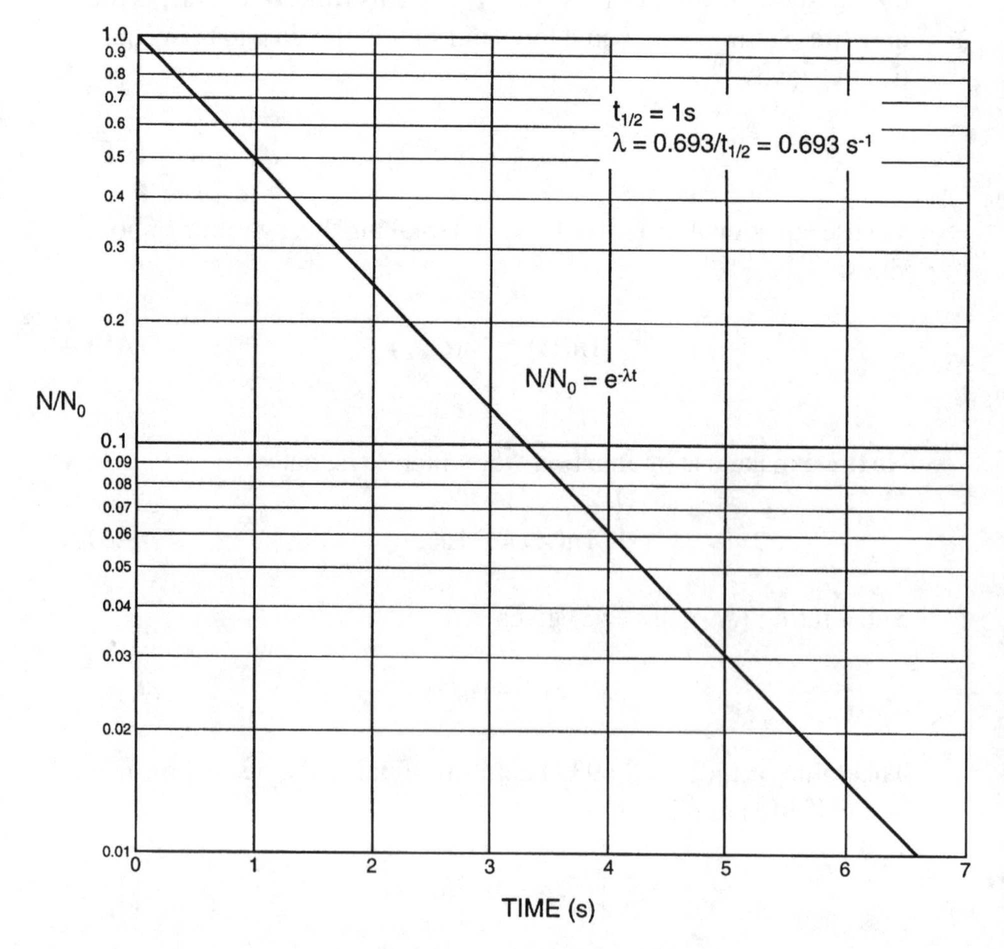

Figure 2.8 Semi-log plot of exponential decay (for a half-life of 1 second).

3

Radiation Applications

With the discovery of the neutron and the advent of nuclear reactors, the large-scale manufacture of useful radionuclides became possible. The number of uses of radionuclides in our modern society is enormous. Most people are familiar with at least a few of the uses in nuclear medicine. Other uses range all the way from food irradiation to carbon-14 dating. These applications of radiation are in addition to x-rays. Some of these uses are described in this chapter. A more complete explanation of radiation applications appears in a delightful book by Dr. John Lenihan titled ***The Good News About Radiation***, published by Cogito Books, Madison, Wisconsin (1993).

NUCLEAR MEDICINE AND RADIOLOGY

Some of the most important uses of radiation and radionuclides are in nuclear medicine and radiology. More than ten million nuclear medicine procedures are conducted each year in the United States to diagnose and treat diseases. One out of every three hospital patients benefits from a nuclear medicine procedure. Radiology is involved primarily with the use of x-rays for diagnosis and therapy. Since Wilhelm Roentgen's discovery of x-rays a century ago, the use of x-rays has continually expanded. Nuclear medicine uses radionuclides for both diagnosis and treatment. Radioactive tracers were first used in 1923 in biological research by the Hungarian chemist Georg Karl Hevesy, for which he received a Nobel prize, and this gradually evolved into the widespread use of radionuclides in medicine. Hevesy has become known as the "father of nuclear medicine." The U.S. scientist Glenn Seaborg and his colleagues at the University of California discovered three radionuclides that have been widely used in nuclear medicine and radiology—technetium-99m, iodine-131, and cobalt-60.

Diagnosis

We are all well aware of the use of x-rays for diagnostic purposes. Enormous advances in this technology have accompanied the spectacular progress in computer technology. Now two-dimensional slices through organs can be reconstructed, and even three-dimensional images can be obtained rapidly by using x-rays together with computer analysis by the technique called ***computed tomography*** (CT), a process invented as late as the 1970s. (You may have heard CT referred to as a CAT scan; this is the same thing, with CAT standing for computer assisted tomography.) I think most people are aware of how important these uses of radiation have been to medical science. More recently the technique of ***magnetic resonance imaging*** (MRI) has been added as a diagnostic tool, though this technique does not use ionizing radiation.

For diagnosis using nuclear medicine, a radionuclide is directed to a particular organ in the body by attaching the radionuclide to a compound that preferentially migrates to that organ. This process of attaching a radionuclide to a compound is called ***labeling***. A ***radiopharmaceutical*** is the combination of a radionuclide and an organ-seeking compound. Some of the principal radiopharmaceuticals, together with the organs for which they are used, are listed in table 3.1. The half-lives of these and several other important radionuclides frequently used in nuclear medicine are given in table 3.2. This list is in no way complete; over forty radionuclides are regularly used. For diagnosis, the radiopharmaceutical is inserted into the body, and the radiation is detected outside the body in a process

TABLE 3.1 Radiopharmaceutical Labeling*

Imaging Objective	*Radiopharmaceutical*
Lung scanning	^{99m}Tc albumin
Bone scanning	^{99m}Tc phosphonates
Gastric mucosa uptake	^{99m}Tc pertechnetate
Thyroid imaging	^{99m}Tc pertechnetate
Liver and spleen imaging	^{99m}Tc sulfur colloid
Bone marrow imaging	^{99m}Tc sulfur colloid
Myocardial (heart) perfusion and muscle imaging	^{201}Tl chloride, ^{99m}Tc isonitriles
Tumor imaging	^{67}Ga citrate, ^{111}In, ^{131}I, and ^{99m}Tc monoclonal antibodies
Imaging of infectious/ inflammatory processes	^{67}Ga citrate, ^{111}In leukocytes
Brain imaging	^{123}I iodoamphetamine, ^{99m}Tc glucoheptonate, ^{99m}Tc DTPA
Renal imaging and function evaluation	^{99m}Tc DTPA, ^{99m}Tc DMSA, ^{131}I hippuran

*From N. P. Alazraki, M.D., and F. S. Mishkin, M.D., ***Fundamentals of Nuclear Medicine***, 2nd Edition, The Society of Nuclear Medicine, New York, 1988.

TABLE 3.2 Half-Lives of Some Radionuclides Used in Nuclear Medicine

Radionuclide	*Half-Life*
^{15}O	2 m
^{13}N	10 m
^{11}C	20 m
^{18}F	2 h
^{99m}Tc	6 h
^{123}I	13 h
^{111}In	2.8 d
^{201}Tl	3.0 d
^{67}Ga	3.3 d
^{131}I	8.1 d

called ***imaging***. The radionuclide used for diagnosis must be a gamma emitter since only gamma photons can escape from the body. The radiopharmaceutical selected for a particular diagnosis must concentrate in the organ of interest, have the right gamma energy, and have a short enough half-life that it will decay away soon after the procedure is completed. Only tiny amounts of radionuclides are needed for most diagnostic procedures.

Radiation from the radionuclide is detected, or imaged, by gamma cameras (also called scintillation cameras) to provide information about the health of the organ. These cameras contain sodium-iodide crystals that give a flash of light (i.e., scintillate) when struck by a gamma ray; these flashes are augmented in photomultiplier tubes and recorded on computers. As is the case with CT scans, two and three-dimensional images of organs can be obtained from gamma-ray tomography with the use of computers. For radiopharmaceuticals that emit single gamma rays, images can be obtained by rotating gamma cameras around the patient; the process is called ***single-photon emission computed tomography*** (SPECT). One of the fastest growing methods in nuclear medicine is ***positron emission tomography*** (PET). As discussed in chapter 2, the positron from a positron emitter is immediately annihilated and produces two 0.51 MeV photons, travelling in opposite directions. In positron emission tomography, a stationary array of cameras is used for detection of the two photons. The two photons must be recorded simultaneously, or in coincidence, to assure that they have been produced by the positron emitter. Positron emitters frequently used in PET include the light radionuclides, carbon-11, nitrogen-13, oxygen-15, and fluorine-18, and the heavier radionuclides, gallium-68 and rubidium-82. C-11, N-13, and O-15 have such short half-lives that they must be produced in a cyclotron located at the hospital. Fluorine-18, Ga-68, and Rb-82 have long enough half-lives that they can be produced elsewhere.

Technetium-99m, with its 140 keV gamma ray, is the most frequently used radionuclide in nuclear medicine. Recently, my foot hurt from playing tennis; the orthopedist thought I had a stress fracture, but it didn't show up in the normal x-ray. So he sent me to a radiologist to obtain a technetium scan, which is more accurate. They put Tc-99m into me and scanned my foot. I didn't have a stress fracture, but that's obviously not the point of the story. The point of the story is what happened when I returned to the nuclear research reactor where I work. As soon as I arrived at the front door, the radiation alarm in the building twenty feet away went off as a result of the radioactive material inside me. Every time I got near the alarm for two days it would go off. On the third day it quit since the half-life of Tc-99m is only 6 hours.

Another time that the alarm went off unexpectedly occurred when we were having a meeting at the reactor with several electric utility executives. When people leave the reactor room after working there or just observing the reactor, they must check their feet and hands on the radiation detector to see if they have picked up any radioactive contamination. In so doing, one of the executives set off the alarm. After washing his shoes and hands, he still set off the alarm. After much consultation, he finally remembered that he had received a heart scan about a month earlier with thallium-201. With its three-day half-life, there was still enough Tl-201 present in his body to set off the detector's alarm.

Treatment

Several methods are used to treat cancer with radiation. One of the most effective methods is the use of high-energy x-rays generated by a ***linear accelerator***. These x-rays are used for the ***external*** treatment of cancer. The linear accelerator is used to accelerate electrons, which then strike a metal target, causing the production of high-energy x-rays called bremsstrahlung (described on page 28). These x-rays are then directed to the cancer tumor. Planning for the geometry of the x-ray beam is aided by such tools as CT and MRI. Sometimes, especially for cancer on or near the skin, electrons are used directly to destroy the tumor. Gamma radiation from Co-60 has also been used for the external irradiation of cancer. In a relatively new machine called a ***gamma knife***, photons from many Co-60 sources are concentrated onto a small space where a tumor is located, thus causing minimum damage to surrounding healthy tissue.

Internal treatment of tumors is performed by a procedure called ***brachytherapy***. In brachytherapy, radioactive materials, often called "seeds," are inserted into the body to kill cancer in a particular organ. Both temporary and permanent implants are used. These radionuclides are generally gamma-ray emitters; if beta particles are also emitted, the seeds must be shielded to prevent excessive local damage by the beta particles. Gold-98, iodine-125, iridium-192, and cesium-137 are all used. Radiopharmaceuticals that concentrate in particular tumors are also used. Iodine-131 is used for thyroid therapy in doses 100 to 2000 times that of diagnostic tracer doses to treat benign and malignant thyroid disease. In the United States 20 000 patients are treated for hyperthyroidism a year by iodine-131, including both President and Mrs. Bush during his presidency. A promising new use of radiopharmaceuticals in cancer therapy is the labeling of monoclonal antibodies that attach only to specific forms of cancer in order to deposit radiation at the cancer site. Iodine-131, yttrium-90, and rhenium-186 are used in this way.

An experimental form of cancer therapy is ***boron neutron capture therapy*** (BNCT). Boron-10 is a strong neutron absorber. When it absorbs a neutron, it produces an energetic alpha particle and a lithium-6 nucleus. In BNCT, a boron-containing drug that selectively concentrates in tumors, is administered to the patient, and the patient is irradiated with neutrons. The energy released in the nuclear reactions with the boron kill the cancer. The method was tried in the 1960s in the United States, but the results were not successful due to the lack of effective boron tumor-seeking drugs. However, new drugs have been invented since then that can deliver 3 to 10 times as much boron to the malignant cells as in the surrounding healthy cells, and considerable success with BNCT has been achieved in Japan for the treatment of previously untreatable brain tumors. Much new research is now underway on the technology in the United States, and the first brain tumor patient was treated in the United States at the Brookhaven National Laboratory Medical Research Reactor in 1994. Only time will tell if BNCT is effective.

Radiation played a big part in the life of my daughter-in-law. Over a decade ago, Kathy developed leukemia and had a bone marrow transplant. She received 200 rads of gamma radiation per day for six days to kill all of the old bone marrow before receiving the new bone marrow. Thus, her total exposure was 1200 rads (also 1200 rems or 12 Sv), which would have been quite lethal without the transplant that followed. Then, during the one to two-month early recuperation period when her immune system was so very weak, she was given only irradiated food to assure the absence of harmful bacteria. For Kathy, the bone marrow transplant was successful, and she is now healthy and vigorous.

FOOD IRRADIATION

Kathy's story leads into a discussion of food irradiation. This process is an extraordinarily effective method of preserving food, and it may become more widely used in the coming years. Sickness caused by harmful bacteria in food is one of the great concerns of the United Nations World Health Organization. In the United States it is estimated that 30 million people get sick each year from microbacterial contamination of food. An estimated 4000 people die in the United States each year from *Salmonella* poisoning, and these problems are more serious worldwide. Other bacteria (or pathogens) found in food beside salmonella—names I can never remember—include *Escherichia coli ("E. coli"), Campylobacter*, *Listeria*, and many others. *E. coli* in hamburgers is the one that has been in the news so much in recent years. Parasites such as trichinae (*Trichinella spiralis*) can also cause food poisoning. Food irradiation kills these microorganisms effectively.

Most irradiated food is irradiated by gamma radiation, mostly from cobalt-60. Some irradiators use cesium-137. Two other radiation sources that can be used are x-rays and electrons from particle accelerators. In any case, the radiation only serves to kill harmful pathogens and parasites. The radiation does not make the food itself radioactive, so a person eating irradiated food does not receive any radiation dose from the food. You know this to be the case with x-rays; getting an x-ray at the doctor's office does not make you radioactive. It's the same way with irradiated food.

Food irradiation has been studied extensively since the 1950s. Irradiation for many food products has been approved by the United States Food and Drug Administration, by the United Nations World Health Organization, and by the United Nations Food and Agriculture Organization. The United Nations group that oversees international food standards, the Codex Alimentarius Commission, found that foods irradiated to 10 kGy present no toxicological hazard, and this commission has approved an international standard that allows irradiation of any food up to 10 kGy.

The amount of radiation allowed in the United States is shown in table 3.3.

High doses in the 20 to 70 kGy range are required for total sterilization, which eliminates all pathogens and parasites. Sterilization would allow the storage of food products, including meat, indefinitely at room temperature. This would allow distribution of food

TABLE 3.3 Doses Permitted by U.S. Food Irradiation Regulations*

Product	*Purpose of Irradiation*	*Dose Permitted (kGy)*	*Date of Regulation*
Wheat & wheat powder	Kill insects	0.2 – 0.5	1963
White potatoes	Extend shelf life	0.05 – 0.15	1965
Spices and dry vegetable seasonings	Decontamination/ Kill insects	30 maximum	1983
Pork	Control trichinae	0.3 minimum to 1.0 maximum	1985
Fresh fruits**	Delay maturation	1	1986
Dry or dehydrated enzyme preparations	Decontamination	10	1986
Dry or dehydrated aromatic vegetable substitutes	Decontamination	30	1986
Poultry	Control illness-causing microorganisms such as salmonella	3	1990

**FDA Consumer*, November 1990.

**Fruits irradiated include strawberries, bananas, avocadoes, mangoes, papaya, guavas, and certain other non-citrus fruits.

that now spoils in many parts of the world before distribution is possible. Radiation-sterilized food is now used in the space program and for certain hospital diets, but general approval has not yet followed. Perhaps someday the U.S. Food and Drug Administration will allow this method of storage.

Despite the potential value of food irradiation to society, opposition to the process has thus far prevented most grocery stores from selling irradiated food. The concern is that irradiation produces some radiolytic products, which opponents fear can lead to harmful effects. Irradiation does produce some organic compounds in food. The amount of these compounds is small, and they are, for the most part, no different from the compounds produced in the normal cooking of food. From reviews of food irradiation research, the U.S. Food and Drug Administration estimates that approximately 90% of the substances identified as radiolytic products are found in foods that have not been irradiated, including raw, heated, and stored foods.

For example, scientists at the U.S. Army Natick Laboratory, where much of the research on irradiated food has been done, found that of 65 substances found in beef irradiated to 50 kGy, only 6 could not be verified in the literature as present in nonirradiated foods, and these six were similar to natural food constituents. Countless animal studies have shown that there are no harmful effects from foods that have been irradiated, although, as with other forms of preservation and cooking, some vitamins are destroyed. Most of the opposition to food irradiation comes from persons outside the technical area of food safety. There is virtually a worldwide scientific consensus among food safety specialists that irradiated food causes no health problems. This becomes quite clear if you attend, as I have done, an international technical conference on food irradiation. It is one of those frustrating situations in which a handful of people can stir up enough fear among the public to prevent society from taking advantage of an extremely beneficial technological advance. Despite this opposition, marketing of irradiated food products is gradually increasing, and its use by astronauts and, when needed, in hospitals continues.

CARBON-14 DATING

An important use of radiation has been its use in dating materials that contain carbon, which includes all organic materials. The method was developed by Willard Libby, for which he received the Nobel prize. Here's how it works.

Most carbon is composed of carbon-12 (C-12), which has 6 protons and 6 neutrons in its nucleus. All carbon in the atmosphere, however, contains some carbon-14 (C-14), which contains 6 protons and 8 neutrons and is radioactive, with a half-life of 5730 years. This C-14 is produced by cosmic radiation. Neutrons in the atmosphere produced from cosmic radiation react with nitrogen-14 in the following neutron-proton reaction to produce C-14:

$$^{1}n + ^{14}N \rightarrow ^{14}C + ^{1}p.$$

These C-14 nuclei react with oxygen to form CO_2 in the atmosphere. An equilibrium concentration of C-14 exists since it is always being produced while simultaneously decaying. This means that there is a constant ratio between C-14 and C-12. This ratio has remained roughly constant for tens of thousands of years.

When plants or animals (or persons) take in carbon from CO_2 or from the food chain, C-14 enters the organism, making all plants and animals radioactive. While these organisms are alive, the carbon in them is constantly changing so that the C-14 and C-12 remain at the same ratio as in the atmosphere. When a plant or animal dies, however, new carbon ceases to enter the organism. After its death, the C-14 begins to decay away with its 5730 year half-life without being replenished. Scientists can measure the ratio of C-14 to C-12 in an organic material and, from this, determine how many half-lives have elapsed since the plant or animal died, thus determining its age. For example, if the relative amount of C-14 is one-half the natural ratio, the material is 5730 years old. If the ratio is one-fourth the natural ratio, the material is two half-lives, or 11 460 years old.

INDUSTRIAL USES

The list of industrial uses of radionuclides is almost endless. I have selected only a few to describe here in the hope that this will give you some idea of the enormous variety of practical uses of the remarkable properties of radiation.

In 1994 the Management Information Services Corporation of Washington, DC, compiled a list of radiation uses that filled 40 pages of their report, ***The Untold Story: Economic and Employment Benefits of the Use of Radioactive Materials***. They claim that, in 1991, radioactive materials in the United States, not counting the nuclear electricity industry, were responsible for $257 billion in total industry sales and 3.7 million jobs.

Radiography. *Radiography* is a nondestructive method of inspection of materials. Just as x-rays are used for medical diagnostics, they are also used to inspect for flaws in metals and other materials, like cracks in metal casings. ***Gamma radiography*** is used in a similar way, for example, to inspect welds in pipelines. In a process called ***autoradiography,*** radionuclides can be incorporated into a material, like a steel casting or a lubricant in a bearing, for its radiographic examination. ***Neutron radiography*** is based on the attenuation of a neutron beam by interaction with materials. This form of radiography is useful to detect the presence of light elements, especially hydrogen, and to measure the presence of elements that absorb neutrons strongly, like boron and cadmium. Neutron radiography is widely used in the aircraft industry to detect corrosion of aircraft components and has been used for the detection of plastic explosives and the measurement of boron concentration in nuclear materials.

Neutron Activation Analysis. *Neutron activation analysis* is a technique to measure trace quantities of impurities in materials. The specimen is irradiated with neutrons, some of which are captured by the trace element. This causes the trace element to become radioactive, and the gamma rays given off by its radioactive decay can be identified (through the gamma energy) with the trace element.

The intensity of the gamma rays is proportional to the concentration of the trace element. An interesting use of this technique is in police forensics; for example, the presence of arsenic in hair or fingernail samples can be verified.

Smoke Detectors. Most smoke detectors are based on the use of radionuclides, especially the alpha emitter americium-241 (433 year half-life). The radioactive source causes a current of charged particles, or ions, to flow in an ion chamber. In the event of a fire, smoke particles enter the chamber, which reduces the ion current. This reduction triggers the smoke alarm.

Product Sterilization. Gamma rays from cobalt-60 are used for the sterilization of a wide variety of items in which the elimination of germs and bacteria is essential. Medical products like surgical dressings, sutures, catheters, and syringes are regularly sterilized with gamma rays. Radiation is used for the sterilization of disposable diapers, tampons, and cosmetics. The possibility of sterilizing food was discussed on page 71.

Radionuclide Gauges. Radionuclides are widely used to measure thicknesses of metals, densities of liquids and chemical systems, and liquid levels in containers all the way from chemical process tanks to beer cans. Use is made of the fact that attenuation of gamma rays is proportional to the thickness or density of a material. Gauges of all kinds are used in a wide variety of industries; radionuclide gauges are both the most accurate and the least expensive type of gauge for countless industrial applications. Most sheet materials produced in large lots are gauged by radiation thickness gauges. Wear and corrosion of metals are also often measured with radioactive materials.

Radioactive Tracers. Minute amounts of radionuclides, called ***tracers***, can be added to materials or systems to follow processes in the materials. Even though the radionuclide may be chemically or physically identical with the other materials in the process, the location of the tracer can be determined from its radioactivity. Flow rates, residence and mixing times, and wear and corrosion in chemical processes are routinely measured with tracers; they are also used for leak detection. The use of radioactive tracers in all fields of scientific research has been used for decades; Hevesy's initial use in biology in 1923 was mentioned above.

Insect Eradication in Agriculture. In addition to food preservation, one of the many uses of radiation in agriculture is in the control of insects in a process called the sterile insect technique. Male insects are sexually sterilized using gamma radiation, and these insects are released into the native insect population. When the sterile male insect mates with the female, no offspring are produced. This technique has resulted in the eradication of a variety of insects. The screwworm, a devastating pest that can kill cattle, has been eradicated in parts of the United States. The Mediterranean fruit fly has been eradicated from Mexico, the Melon fruit fly from Okinawa, and several fruit flies from the United States. One specie of tsetse fly has been eradicated from parts of Africa. The technique is currently being tried for the control of the gypsy moth in the United States.

Elimination of Static Electricity. Static electricity is a hazard for some moving machinery; it can cause fires, explosions, and production interruptions. When grounding of equipment is inadequate to solve the problem, radiation can be used to ionize the air where charges accumulate. This neutralizes the electrostatic charge and prevents its buildup. This method of controlling static electricity is used in the printing and paper industries.

Vulcanization of Rubber and Curing of Plastics. Radiation is one of the methods used in the automobile tire industry for the vulcanization of rubber. Both cobalt-60 and electron beam radiation are used to cure polymers (that is, to "cross-link" the long chains of organic molecules), like polyethylene, used for high-temperature or long-lasting, high-reliability insulation on electrical cables used in nuclear power plant safety equipment and in airplanes.

4

Fission and History

FISSION

Fission is a special nuclear reaction. It is the source of energy in a nuclear power plant. Fission results from a reaction between a neutron and the nucleus of a heavy element, like uranium or plutonium. A quite different nuclear reaction that produces energy is ***fusion***, which is the subject of chapter 10. Fusion involves light elements, like the isotopes of hydrogen.

In a fission reaction, the heavy nucleus splits into two fragments, called ***fission products***, and a relatively large amount of energy is released. In addition, two or three neutrons are ejected. The process is illustrated in figure 4.1. We have to be careful here. I said "relatively" large. The absolute energy of a single fission event is quite small since we are talking about objects the size of an individual atom, or nucleus. However, compared to the energy released in a chemical reaction (as in the normal burning of coal) involving a single atom or molecule, the fission of a single uranium atom is enormous—millions of times as large. Thus, the energy from the fissioning of a gram of uranium releases millions of times more energy than the burning of a gram of coal. The unit of energy that we generally use to describe nuclear reactions, the MeV, was discussed in chapter 2. The energy released in the fission of a single uranium

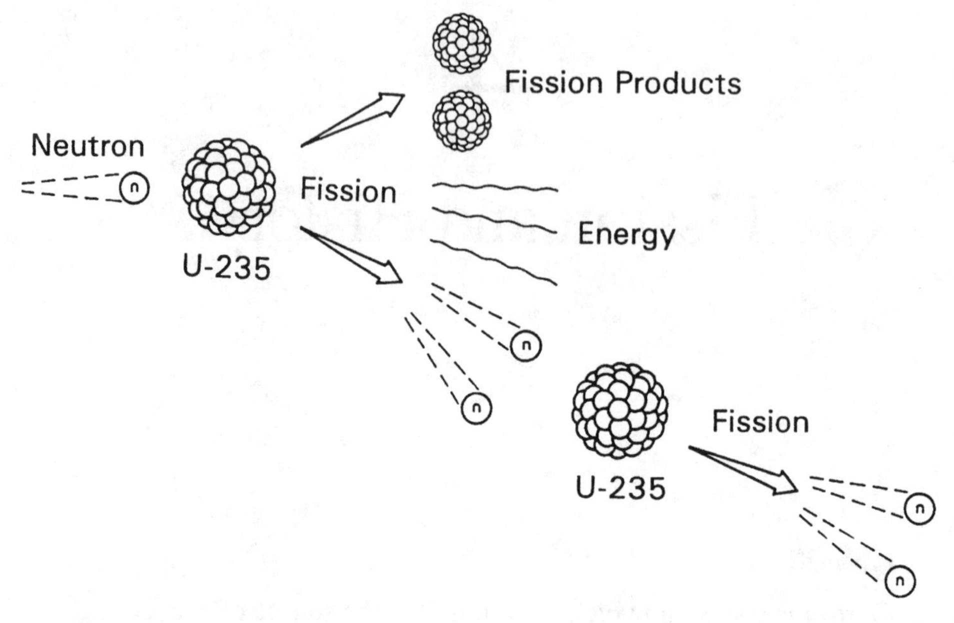

Figure 4.1 Nuclear fission. COURTESY OF DOE.

nucleus is about 200 MeV. By comparison, the energy released in the chemical burning of a single atom of carbon (as in coal) is about 3 eV, a factor of 70 million lower than the energy per fission.

All heavy elements can undergo fission if they are struck by neutrons that have sufficient energy. Several isotopes of uranium and plutonium have a unique property; they will fission when struck by neutrons that have very little energy. These low-energy neutrons are called ***thermal neutrons*** because they have kinetic energies of the same order of magnitude as atoms and molecules at ordinary room temperature, around 0.02 eV. This thermal energy is very low compared with the high-energy particles that physicists so often study. The most important of these isotopes that undergo fission by thermal neutrons is uranium-235. The next most important one is plutonium-239. Others include uranium-233 and plutonium-241. These isotopes have a generic name; they are called ***fissile isotopes***. The probability that a low-energy neutron will cause fission when it collides with U-235 or one of the other fissile isotopes is much higher than the probability of fission from a high-energy neutron. Fission of U-238 can occur by collision with high-energy neutrons (of the order of 1 MeV and higher) but not by collision with thermal neutrons.

In most (though not all) nuclear reactor designs there are a lot of thermal neutrons moving about (to take advantage of the high probability of fission by thermal neutrons). These reactors are called ***thermal reactors***. When I said a lot, I meant a whole lot, like a billion thermal neutrons per cubic centimetre in a thermal reactor. Since we want fissions to occur prolifically in a reactor to make energy, U-235 is an excellent candidate for fuel for thermal reactors. So would be the other fissile isotopes, but U-235 is the only one that is abundant in nature. All U-233, Pu-239, and Pu-241 is artificial; most that does exist was made in nuclear reactors.

Uranium-235 is found in all natural uranium ore. However, only 0.7% of natural uranium is U-235. Almost all of the rest—the other 99.3% of it—is U-238. Now you see why I keep harping on these two isotopes. Thus, regrettably, only a small fraction of the uranium mined is very useful for current commercial reactor designs.

When a neutron is absorbed by a uranium nucleus and causes that nucleus to fission, several things happen. We have already referred to the release of a large amount of energy and to the breaking apart of the nucleus into two fission fragments, or fission products. Also shown on figure 4.1 is the production, or ejection, of several neutrons from the uranium nucleus. On the average, 2.44 neutrons are produced in the thermal fission of U-235. The energy of the neutrons produced in fission is very high, in the range of 0.5 MeV to 14 MeV. The average energy of the neutrons produced is 2 MeV, which is 100 million times the kinetic energy of the thermal neutron that caused the fission. Some gamma radiation is also emitted at the instant of fission.

The fission products are radioactive. They decay with the emission of both beta particles and gamma rays. Most of these decay rapidly, but some have very long half-lives. These constitute most of the high-level waste from nuclear power plant operation, which is the subject of chapter 8. Some of the most important fission products were discussed in chapter 2, including strontium-90, cesium-137, iodine-131, and the very long-lived isotopes technetium-99 and iodine-129.

Energy Generated in Fission

The next question to ponder is, what is the form of this large amount of energy released in fission? Most of it, about 85%, is initially the kinetic energy of the fission products—just from the initial high velocity of these products as they erupt from the fission site. You may recall that kinetic energy is one-half the product of the particle's mass and the square of its velocity ($\frac{1}{2}mv^2$). These fission products slow down almost instantaneously by colliding with uranium and other atoms in the fuel, thus transferring their kinetic energy to the fuel. It is this energy that heats up the fuel, thereby providing the thermal energy, or heat, needed to make the nuclear power plant work.

Where does the other 15% of the energy come from? Some comes from the gamma rays that were emitted during the fission process. Some comes from the kinetic energy of the neutrons that were ejected. Some of the energy is ***delayed***. This means that it appears some time after the actual fission event. The energy released instantaneously at the time of the fission is called ***prompt***. The delayed energy comes from the radioactive decay of the fission products. This delayed energy plus the prompt gamma rays and the neutron kinetic energy constitute the other 15% of the fission energy. Like the energy from the fission fragments, these forms of radiation are captured by the fuel (and by other materials adjacent to the fuel) to produce heat that is used by the power plant. Hence, all of this energy is useful.

This concept of delayed energy from the fission products is really a very important aspect of nuclear reactor safety, as we shall see later; so don't forget about it.

As is so often the case in science, the situation is actually a little more complicated than what I've just described, and two added complexities are worth mentioning here. As a good nuclear engineer interested in the application of nuclear fission, I am most interested in the energy that is useful in producing heat, as is all the energy discussed above. In reality a little more energy is released during fission (about 5% more) in the form of ***neutrinos***. We engineers don't talk about neutrinos much because they escape completely from the power plant; they are utterly useless and don't cause any harm to anyone anywhere, and I can't even tell you where they go—to outer space or the middle of the Earth, I suppose! You may recall that I did talk about them in chapter 2 when I discussed beta particle energies. While we engineers don't care much about neutrinos, physicists are interested in them because physicists are compelled to account for the total energy released in fission, and neutrinos are part of that total. The second minor complexity is that still more energy (which I haven't talked about yet) is eventually produced by some of the neutrons that were created in fission. This comes from the fact that some neutrons in a nuclear reactor are ***captured*** by materials that do not result in fission, and a little energy comes from this neutron-capture process in the form of gamma rays. This is useful energy that design engineers must account for. The amount of energy from neutron capture is about the same as the neutrino energy that is lost.

The energy generated by fission is summarized in table 4.1 for a typical thermal reactor.

TABLE 4.1 Energy Per Fission*

	MeV
Kinetic energy of fission products	165
Kinetic energy of fission neutrons	5
Gamma rays produced instantaneously	7
Gamma rays from fission products	6
Beta particles from fission products	7
Neutrinos	10
TOTAL	200 MeV

* For useful energy, one must add about 10 MeV per fission due to capture of neutrons and delete the neutrino energy, so that the useful energy also comes out to be about 200 MeV.

Conversion of Mass into Energy in Fission

Now let's see how the physicists account for every bit of the energy released. The energy released in fission literally comes from the conversion of mass into energy. You start with Einstein's remarkable equation,

$$E = mc^2$$

which was discussed in chapter 2. For a U-235 nucleus undergoing fission by a thermal neutron, the initial mass involved in the nuclear reaction is the combined masses of the U-235 nucleus and the neutron that causes the fission. The final mass is the mass of the fission products plus the mass of the neutrons and beta particles released. The difference between the original and final masses is exactly the energy released during fission, as calculated from $E = mc^2$.

Let me illustrate with an example. Suppose a thermal neutron causes fission in U-235 and the resulting fission products are strontium-94 (Z = 38) and xenon-139 (Z = 54) plus three neutrons. (Z = the atomic number, see page 18.) The strontium-94 then decays to stable zirconium-94 (Z = 40) by emission of two beta particles, and the xenon-139 decays to stable lanthanum-139 (Z = 57) by emission of three betas. Neutrinos and gamma rays are also emitted, but they have neither mass nor charge and are not included in our mass calculation. The overall nuclear reaction, including the beta particles, is illustrated by the equation:

$${}^{1}_{0}\text{n} + {}^{235}_{92}\text{U} \rightarrow {}^{94}_{40}\text{Zr} + {}^{139}_{57}\text{La} + 3\,{}^{1}_{0}\text{n} + 5 \text{ betas.}$$

By doing some adding, you can observe that the total number of neutrons plus protons and the total charge have been conserved.

Next compare the masses of the reactants and the products, which are known with great accuracy. We use the atomic mass unit (amu).

neutron	1.008 665 amu
U-235	235.043 91 amu
Zr-94	93.906 31 amu

La-139	138.906 14 amu
beta	0.000 55 amu (about 1/1840)

Thus the total mass on the left hand side of the equation (the reactants) is

$$1.008\ 665 + 235.043\ 91 = 236.052\ 58 \text{ amu.}$$

The total mass on the right hand side (the products) is

$$93.906\ 31 + 138.906\ 14 + (3 \times 1.008\ 665) + (5 \times 0.000\ 55) = 235.841\ 20 \text{ amu.}$$

There is definitely a difference between the masses of the reactants and the products—not large, but not zero. The difference is 0.21138 amu. Mass has very clearly not been conserved (which is different from what we were taught for chemical reactions).

Recall that one amu is equivalent to $1.660\ 56 \times 10^{-27}$ kilograms. Therefore, the difference of 0.211 38 amu is equal to 3.510×10^{-28} kg. This difference is the mass that has been converted into energy. Finally, using the conversion factor of 1.602×10^{-13} J/MeV and recalling that $c = 2.998 \times 10^{8}$ m/s, we make use of Einstein's equation to calculate the energy release in our example fission:

$$E = (3.510 \times 10^{-28} \text{ kg}) \times (2.998 \times 10^{8} \text{ m/s})^2 / 1.602 \times 10^{-13} \text{J/MeV} = 197 \text{ MeV},$$

and voila, this is the energy released for this particular fission event.

We said earlier that the ***average*** energy per fission was about 200 MeV. A fission event can produce a variety of fission product pairs and number of neutrons and betas different from our example and, hence, energies different from our calculated 197 MeV. 200 MeV is the average of all of the different reactions that occur in the fission of U-235.

This energy is precisely the energy released in fission when the neutrinos are included and when energy released in the capture of neutrons is not. As you can see, it comes directly from $E = mc^2$. Now, to me, that's amazing. Not incredible, because I believe it, absolutely. But amazing. Mass is indeed converted into energy in the fission process.

Sometimes the following conversion factor is useful: 1 amu = 931 MeV. This conversion equates mass and energy directly through Einstein's equation. Thus, multiplying the mass difference of 0.211 38 amu by 931 MeV will give us 197 MeV directly.

UN PETIT DIVERTISSEMENT INTO HISTORY

How did the world learn about this marvelous phenomenon that we call fission and about radiation and the conversion of mass into energy? The story is fascinating, sufficiently so that this is a good time to leave technology for a while and explore some history.

Perhaps we might take as a starting point the principle of conservation of mass in chemical reactions that was so elegantly proved near the end of the eighteenth century by the great French chemist Antoine-Laurent Lavoisier. Unfortunately, this noble scientist eventually became a victim of the guillotine in the French Revolution. After Lavoisier came the remarkable atomic theory of the English physicist John Dalton (1808) in which he proposed that elements are composed of tiny indestructible particles called atoms and that all the atoms of a particular element are alike. This was followed in 1811 by the law formulated by the Italian Amadeo Avogadro that equal volumes of different gases at the same pressure and temperature contain the same number of molecules. The number of molecules in a mole of material bears Avogadro's name.

Then there followed the theory of the conservation of energy. Actually, the Dutch physicist Christiaan Huygens, a contemporary of Newton and friend of Leibniz, as long ago as 1669, argued that the product of mass and the square of the velocity was conserved upon the impact of two bodies, which we now know as the conservation of kinetic energy. It is hard for us to realize that as late as the eighteenth century, heat was thought to be some fluid that passed from one body to another during a heat transfer process. It was at the end of that century that Benjamin Thompson (Count Rumford) showed, in 1798, that heat was not a fluid but a form of energy. By 1840 James Joule was measuring the conversion of mechanical energy into energy transferred as heat, and soon there followed the proof of the conservation of energy. The recognition of the equivalence of mass and energy, however, was still a half-century away.

For our saga the next dramatic event regarding mass and energy was the theory of their equivalence, which the great German physicist Albert Einstein reported to the world in one of four dramatic papers he published in 1905. The other three included an explanation of Brownian motion, an explanation of the photoelectric effect, and his special theory of relativity! His theory on mass and energy was later verified by studies of nuclear reactions, but it was to be more than thirty years before scientists discovered the awesome release of energy from the conversion of mass into energy in the fission process.

Shortly before the end of the last century, in 1895 in Bavaria, Wilhelm Roentgen was experimenting with a gas discharge tube. The tube was covered with black paper, and the room was dark. Roentgen noticed that a crystal some distance from the tube was luminescent, glowing, whenever the tube voltage was on. Something was being transmitted across the space between the tube and the crystal. Roentgen had discovered x-rays. This momentous discovery was followed the next year by the accidental discovery by Henri Becquerel, a Frenchman, that uranium produced an image on a photographic plate. Marie and Pierre Curie figured out what was happening, and Madame Curie gave the name radioactivity to the phenomenon. Becquerel and Marie and Pierre Curie jointly received the Nobel Prize in physics in 1903 for the discovery of radioactivity.

Marie Curie was from Poland and went to France to study, where she met and married Pierre Curie. They recognized that pitchblende ore from which uranium was extracted was more radioactive than uranium itself, which led them to believe that there must be some radioactive element in the ore that was more radioactive than uranium. Together, in 1898, they announced the discovery of the element polonium, named for her native country, and later that winter Marie Curie had also isolated radium from pitchblende. For her discovery of radium Marie Curie was awarded a second Nobel Prize, this time in chemistry. Today we accept the fact that women's contributions to science and engineering equal men's contributions. This was not the way it was, however, when I was in college; imagine, Madame Curie made her discoveries 100 years ago! Surely this is one of the most remarkable contributions by a single person in human history. And miraculously, Madame Curie's daughter, Irène Joliot-Curie, together with her husband Frédéric Joliot-Curie, received the Nobel Prize for chemistry in 1935 for the first production of artificial radioactivity. (Another curiosity—hyphenated married names way back in 1935! I've been told that this was to carry on the Curie name since Marie and Pierre Curie had daughters only.)

Soon after radioactivity was discovered, Ernest Rutherford, from New Zealand but working in England, identified the radiation from uranium as two types of radiation, which he called alpha rays and beta rays. He showed that one element could change by alpha or beta decay into a completely different element, which brought out great skepticism among chemists—who, of course, disbelieved in alchemy. Around 1900 Rutherford's predecessor as the head of the Cavendish Laboratory at Cambridge, J.J. Thomson, discovered the electron. Finally, it was Rutherford, in 1911, who discovered the nucleus—that nearly all of the mass and all of the positive charge was concentrated in the center of the atom. Rutherford's principal colleague for his famous experiment was Hans Geiger, inventor of the Geiger counter. I recall that three students and I had the fun of repeating Rutherford's famous experiment in a semester-long project during my senior year in college. Rutherford's momentous discovery was followed in 1913 by the exposition of the atomic model of electrons orbiting around the nucleus by the Danish physicist Niels Bohr. For the few who understood what was going on, those must have been exciting times.

The remarkable discoveries of the 1890s—x-rays by Roentgen, radioactivity by Becquerel, and radium by the Curies—led to the entirely new fields of medical radiology and eventually radiation protection and health physics. The use of Roentgen's x-ray tube expanded rapidly, both for its diagnostic capability and its ability to fight cancer. It was not long before health technicians became aware of the dangers of x-rays. X-ray burns caused skin to redden. That x-ray damage deeper in the body could lead to cancer was not recognized until some time later.

Radium produced gamma rays, which were higher in energy than most x-rays. The higher energy radiation from radium also became a useful tool in combating disease, especially cancer, so radium production accelerated. The radium isolated by Marie Curie was radium-226. Later, a cheaper form of radium, radium-228, called mesothorium, was used. With the use of radium, a new danger soon became evident. Unlike the source of x-rays, radium was a material that could enter the body and become a poison.

One non-therapeutic use of radium in the first four decades of the twentieth century was for luminous watch dials. A tiny amount of radium was mixed with zinc sulfide to make the sulfide fluoresce. Radium dials were painted by hand, mostly by women in factories in New Jersey. The practice was to "tip" the brush by whetting it with the tongue to produce a fine point for painting the dial. In the process, tiny quantities of radium entered the bodies of the painters. In 1924, a dial painter was found to have cancer in her jaw from the radium. More radium dial painters died during the next few years. The question arose, how much radium could one tolerate in the body without harm? The greatest researcher in this field was Robley Evans, a physicist at MIT. He investigated the dial painters to find an answer to this question.

Two decades after Evans's first discoveries, during my senior year in college, I had the wonderful opportunity to take Professor Evans's course in nuclear physics, and even the better opportunity to have him as my advisor for my senior thesis project. I measured mesothorium (radium-228) concentrations using a photographic method that Evans's lab was developing in which each alpha decay could be observed in a film under a microscope. One alpha decay would produce a track that we called a "one-track star," two alpha decays would produce two tracks starting from the same point that we called "two-track stars," and so on as more alphas were emitted.

Mesothorium eventually decays to lead through the emission of five alpha particles. In a few cases I actually saw "five-track stars." This meant that in the few days between the time that I put a mesothorium solution onto the photographic emulsion and the time that I developed the film, a few mesothorium nuclei decayed all the way to lead. The original mesothorium concentration could be calculated from the number of different stars. What an excitement that was to me, to observe so convincingly the statistical nature of radioactivity. I actually worked mostly with a graduate student of Professor Evans's named Norman Rasmussen, who later became head of MIT's Nuclear Engineering Department and the author of the most famous study that has ever been made in nuclear reactor safety, report number WASH 1400, more widely known as "the Rasmussen Report."

This history of radiation ***health physics*** (as it was later called) is well chronicled in Barton Hacker's book, ***The Dragon's Tail*** (University of California Press, 1987). The dangers from x-rays and radium became well enough known by the late 1920s that international radiation committees were formed for their regulation, and the first standards were developed in 1928. The unit of dose for radiation exposure was an unusually difficult parameter to define and measure, and it is remarkable that the fundamental unit of radiation exposure, the roentgen, was not widely adopted until 1928, three decades after Roentgen's historic discovery. This unit was followed by the rem and the rad, and later by the SI units, the sievert and the gray, which are defined in chapter 2. It's hard to believe now that the first widely used unit of dose was the "unit skin dose," or "erythema dose," which was the amount of x radiation that caused reddening of the skin! An early practice of radiologists was to limit their monthly exposure to 1/100th of a unit skin dose. Fortunately, the roentgen eventually came along, which was based on ionization in air rather than a biological effect, and therefore could be accurately measured.

The neutron was discovered in 1932 by James Chadwick at Cambridge University. Within months after the discovery of the neutron, a Hungarian physicist named Leo Szilard (who was later to play an important role in motivating the United States to begin development of the atomic bomb at the start of the Second World War) actually took out a patent on a device that would exploit the enormous energy of the nucleus using a chain reaction based on a neutron-capture reaction involving the release of more than two neutrons, which is, we now know, what a nuclear reactor does. At the time Szilard had no idea whether such a reaction could ever happen; fissioning of uranium had yet to be discovered. In the 1930s, Enrico Fermi developed the theory of neutron diffusion and slowing down in a medium like graphite or water and applied his theories to experiments to make radioactive materials from the absorption of thermal neutrons. Fortunately for science and the Allied cause in the Second World War, Fermi was awarded the Nobel prize in physics in 1938 for his work on thermal neutrons. This allowed him to travel to Sweden and escape from the fascist government that ruled Italy at that time. Also during this decade, Ernest O. Lawrence of the University of California invented the cyclotron, which was used for the discovery of elements heavier than uranium, called ***transuranic elements***. The most important was plutonium, discovered in 1941 by Glenn Seaborg, a chemist at the University of California.

Then came the most puzzling discovery of all for our story. In 1938 two German scientists, Otto Hahn and Fritz Strassmann, reported (with great hesitation) that upon bombarding uranium with neutrons, the much lighter element barium was produced. No matter how careful they were, they always obtained barium, which made no sense at all. Lise Meitner was a colleague of Hahn and Strassmann in their landmark experiments, but she was a Jew and escaped from Germany to Sweden to flee the Nazi persecution of the Jews. There she, together with her nephew Otto Frisch, formulated the explanation that the uranium nucleus had split into two products, of which barium was an example. Such was the discovery of fission.

The recognition of the tremendous potential for producing energy from fission, along with the potential for making weapons beyond imagination, was immediately recognized by the world's nuclear physicists, who grasped what Hahn, Strassmann, and Meitner had done. In 1939, on the eve of the Second World War, Szilard and Eugene Wigner convinced Einstein to write a letter to President Franklin D. Roosevelt about the importance of the new discoveries. This led to a meeting of scientists with President Roosevelt, and he established a committee to direct activities on fission for the next few years. In 1942 Roosevelt established the Manhattan Project to develop the atomic bomb, in complete secrecy.

Before the Manhattan Project, the main group of scientists working on fission were brought together at the University of Chicago under the leadership of Arthur Compton (of Compton effect fame). The first controlled nuclear reactor was assembled in secret, under Fermi's direction, on a squash court beneath the stands of Stagg Field at the University of Chicago in the fall of 1942. This was a graphite "pile" (as it was called then) in which graphite blocks were stacked and fuel in the form of natural uranium was inserted between the graphite blocks. A drawing of the pile is shown in figure 4.2. The objective was to obtain a stable ***chain reaction*** in which the neutron level could be maintained from fission without the presence of an external source of neutrons. A reactor in this condition is said to be ***critical***, or to have achieved ***criticality***. A ***critical mass*** is the mass of uranium needed in a particular geometry to achieve criticality. (The concepts of criticality and chain reaction are described in further detail in chapter 5.) Natural uranium can form a critical mass when placed in a graphite or a heavy water environment—and only in the presence of these two materials, which have just the right properties of not absorbing too many neutrons and being low in molecular weight. These low molecular weight materials in a nuclear reactor are called ***moderators***, whose function is described in chapter 5. The experiment was a success, and the world's first self-sustaining fission chain reaction took place on 2 December 1942. A full history of this first chain reaction, together with the first fifty years of history of nuclear energy development, appears in the book ***Controlled Nuclear Chain Reaction, the First 50 Years***, American Nuclear Society, 1992.

Figure 4.2 Drawing of the 1942 pile on the Stagg Field squash court. COURTESY OF ARGONNE NATIONAL LABORATORY.

An early task of the Manhattan Project was the construction of an air-cooled natural uranium reactor in Oak Ridge, Tennessee, in 1943. Later to become the Oak Ridge National Laboratory, the lab was called the Clinton Engineering Works at the time, named after the nearby town of Clinton. Before the war, the Oak Ridge area was completely rural. This early reactor led to the construction of production reactors in the eastern Washington desert, at Hanford, where the plutonium for the first atomic bombs was made. The development of the atomic bomb is a fascinating story. It is too bad, but true, that the events that announced to the world the advent of nuclear energy were the explosions at Hiroshima and Nagasaki. In addition to the secret laboratories at Oak Ridge and Hanford, laboratories at Los Alamos, atop a mesa in northern New Mexico, and at Argonne, near Chicago, were built to develop the bomb.

Many of the world's leading scientists were assembled at Los Alamos under the direction of J. Robert Oppenheimer to figure out how to make a bomb. They made a uranium bomb from uranium-235 and a plutonium bomb, composed mostly of the isotope plutonium-239. The U-235 bomb was a gun-type device in which two halves were shoved together to make it explode. The plutonium bomb was triggered by an implosion device, in which chemical explosives surrounding a sphere of plutonium drove the plutonium together, actually compressing the metal into a critical mass. In a project called Trinity, a plutonium bomb was secretly detonated on 16 July 1945 in a highly successful test at Alamogordo in southern New Mexico, now the White Sands Missile Range. Imagine—only two and a half years after Fermi's reactor reached criticality.

The Hiroshima bomb was made of uranium-235. It was a gun-like design, which had never been tested. The uranium-235 was obtained from Oak Ridge, from a process that used huge mass spectrographs called "calutrons." The Nagasaki bomb used plutonium obtained from plutonium production reactors at Hanford.

After the war, engineers began to develop methods to use nuclear energy for propulsion and the generation of electricity. On 20 December 1951, at a remote site in Idaho, a team of engineers led by Walter Zinn threw a switch that lit a string of four electric lightbulbs from a generator powered by a ***fast reactor***, called the Experimental Breeder Reactor (EBR-I). This was the first electricity ever produced with nuclear energy. This was followed by the advent of the nuclear submarine and the nuclear navy under the leadership of Admiral Hyman Rickover. The nuclear submarine used a ***pressurized water reactor*** (PWR), and this reactor was soon adopted for commercial nuclear energy. The first nuclear power plant to go onto the commercial electric grid was Shippingport, a 150 MWe reactor built in Pennsylvania by the federal government's Atomic Energy Commission and Rickover's navy organization. The first fully commercial PWR was designed by Westinghouse and built for the Yankee Atomic Power Co. at Rowe, Massachusetts; it began producing 250 MWe in 1958 and operated until 1991. In the meantime, Sam Untermeyer and his colleagues at the Argonne National Laboratory invented the ***boiling water reactor*** (BWR), and in 1960, Commonwealth Edison started up the first commercial BWR at Dresden, Illinois, a 250 MWe BWR designed by General Electric.

Since that time more than a hundred commercial plants have operated in the United States, and over four hundred throughout the world in more than 25 countries, in sizes up to 1450 MWe. In the early 1990s between 15 and 20% of the world's electricity was produced by nuclear energy. Nearly all of this electricity comes from PWRs and BWRs. In the United States, the PWRs were built by Westinghouse, Combustion Engineering, and Babcock & Wilcox. The BWRs were built by General Electric. The Canadians use a different type of pressurized water reactor called ***CANDU,*** which uses heavy water as the coolant and moderator. In the former Soviet Union, both PWRs and another type of water-cooled graphite reactor called ***RBMK*** are used. ***Liquid metal (sodium) breeder reactors*** (LMR) have also been built, but they are not yet economically competitive. While the United States pioneered the breeder reactor, France and the USSR took the leadership in this development in the 1970s and 80s. The United States reentered the LMR picture in the late 1980s with a new design described in chapter 7. ***Gas-cooled reactors*** were also operated early in the development of commercial nuclear power, especially in the United Kingdom, but only recently have new designs been developed that stand a chance of becoming economically competitive. These new designs are also discussed in chapter 7.

Well, so much for history. Scientific history and the amazing discoveries that got us to where we are in the nuclear energy field fascinate me. But now, back to applied science and engineering and how a nuclear power plant works.

5

How a Nuclear Power Plant Works

A nuclear power plant consists primarily of a nuclear reactor inside a containment building, alongside a turbine/generator building. This is illustrated in figure 5.1. The reactor produces energy in the form of heat, which is used to generate steam. This steam drives a turbine/generator system where the electricity is made. A generator consists of magnetic poles between which coils of wire rotate to produce electric current—one of the marvels of the physical world. The whole objective of an electric power generating station is to rotate the coils in this generator to produce electricity. A turbine is used to rotate a shaft, which rotates the coils in the generator. Thus any power plant—whether it be nuclear, fossil, or hydroelectric—is designed with the common goal of turning a turbine to run a generator. The details of this process in a nuclear plant are described in this chapter.

A typical large nuclear power plant generates electricity at the rate of about 1000 MW. We label this 1000 MWe in order to distinguish between electricity and thermal energy (heat) generation. A 1000 MWe plant will produce heat at a rate of about 3000 MW, which we label MWt.

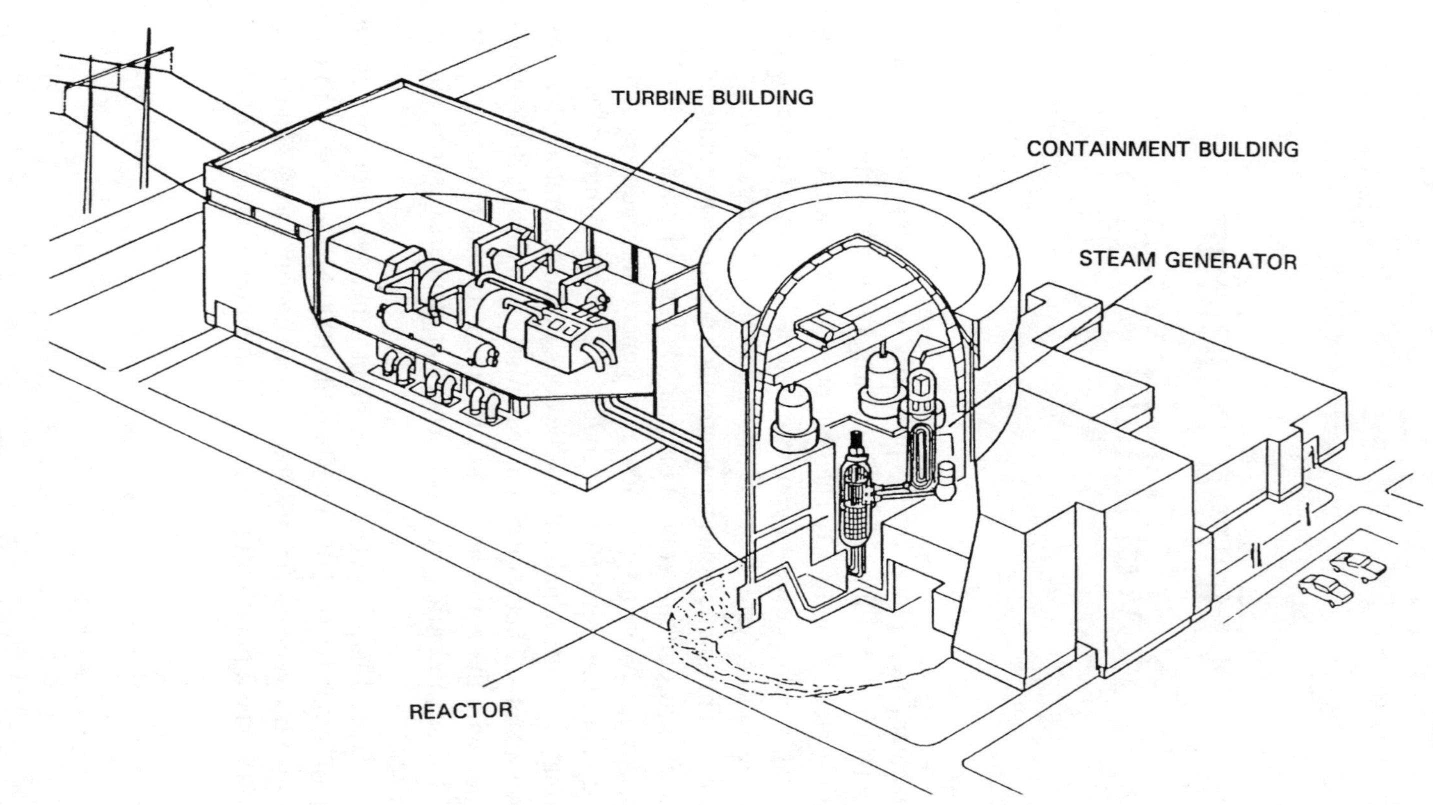

Figure 5.1 A nuclear power plant.

THE NUCLEAR REACTOR

The central part of a nuclear power plant is the ***nuclear reactor***. The nuclear fuel is assembled in the ***reactor core***, and it is here where the fission takes place.

There are several different types of nuclear reactors used for commercial power generation, and several more in the design and development stage. I will spend most of this chapter on the ***light water reactor*** (LWR), which is the main type used throughout the world and the only type in commercial operation at this time in the United States. We say "light water" (ordinary H_2O) to distinguish from "heavy water," which is composed of D_2O instead of H_2O. Heavy water is used in the CANDU reactor manufactured in Canada; this reactor is described near the end of this chapter.

There are two kinds of commercial light water reactors—the ***pressurized water reactor*** (PWR) and the ***boiling water reactor*** (BWR). In both cases water flows past uranium fuel to remove heat from the fuel, or to cool the fuel. Hot water is needed to make electricity in a steam cycle electric power plant, whether the plant be nuclear, coal, gas, or oil. Shortly we will see the role of water and steam in this process. But for now let's return to the reactor core, the uranium, and the fission process. This is the part we are most interested in—the uniquely nuclear part.

The fuel in a PWR and in a BWR are nearly identical, and the reactor cores are similar. The fuel for both is uranium dioxide, UO_2. This is a ceramic material, with a crystalline structure. It is hard and black. It is produced by ***sintering***, which means firing at a high temperature. The fuel is made into small pellets, about 1/3 inch in diameter and 1/2 to 3/4 inches long, as illustrated in figure 5.2. The steps involved in the assembly of the fuel into the reactor are illustrated in figure 5.3. The UO_2 pellets are stacked in metal tubes, called ***cladding***, about 250 pellets per fuel rod. The tubes are made of an alloy of the metal zirconium, called zircaloy. (It's always been a mystery to me why zircaloy has only one l, but some things I just accept.) The outside diameter of these tubes is a little less than half an inch. The tubes are 12 feet (3.7 m) long. A tube filled with fuel pellets is called a ***fuel rod***, or sometimes a ***fuel pin***.

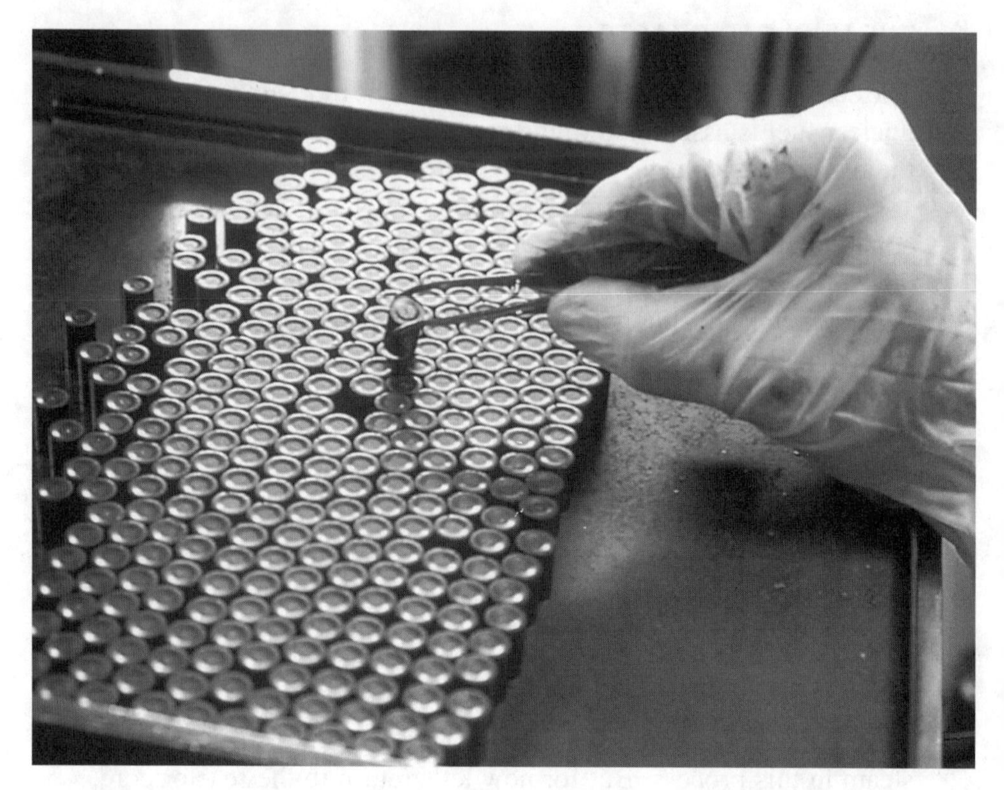

Figure 5.2 Uranium dioxide fuel pellets. COURTESY OF FRAMATOME COGEMA FUELS

These fuel rods are next assembled into what is appropriately called a ***fuel assembly***. The rods are assembled in a ***square lattice***. A fuel assembly for a PWR is illustrated in figure 5.4 and for a BWR in figure 5.5. The PWR assembly might contain anywhere from 15 by 15 rods to 17 by 17 rods. The BWR assembly contains only about 8 by 8 rods, and it has a zircaloy shroud, or can, or wrapper, around it. The repeating lattice of fuel assemblies for a BWR is illustrated in figure 5.6.

The fuel assemblies are assembled in the reactor core. The core is located in a large pressurized vessel that contains the water that cools the fuel rods. The PWR core and reactor vessel are illustrated in figure 5.7; the core contains between 200 and 300 fuel assemblies. A BWR core and vessel are shown in figure 5.8. The top view of a

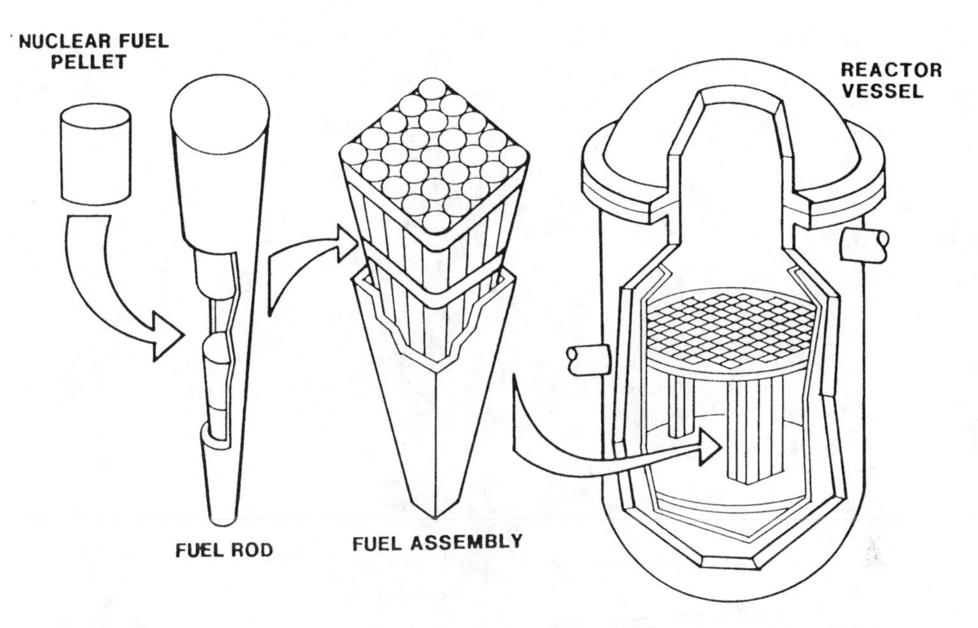

Figure 5.3 Assembly of the fuel into a core. Courtesy of the American Nuclear Society.

BWR core containing 956 fuel assemblies is shown in figure 5.9. (A portion of figure 5.9 appeared in figure 5.6.) The crosses in figure 5.9 represent control rods, which are described later. The shape of both a PWR and a BWR core approaches that of a large cylinder, with a diameter of about 12 feet (3.7 m).

In a PWR the pressure inside the reactor vessel is 2250 ***pounds per square inch (psi)***.* This is high pressure. Atmospheric pressure is 14.7 psi. Thus the pressure in a PWR is 153 atmospheres.

---

*I will quote most pressures in ***psi*** since this is what is most frequently used by the U.S. nuclear and electric utility industry. It is useful to begin to become acquainted with the SI unit of pressure, the ***pascal*** (Pa), so sometimes I will also give pressures in this unit. Actually, we will use ***megapascals*** (MPa), which is 10^6 Pa, since a pascal is so small. I will also quote pressures in ***bars*** since one bar is almost equal to one ***atmosphere*** (which is the normal air pressure in the atmosphere at sea level) and people will generally have some "feel" for an atmosphere. One bar is also equal to 0.10 MPa.

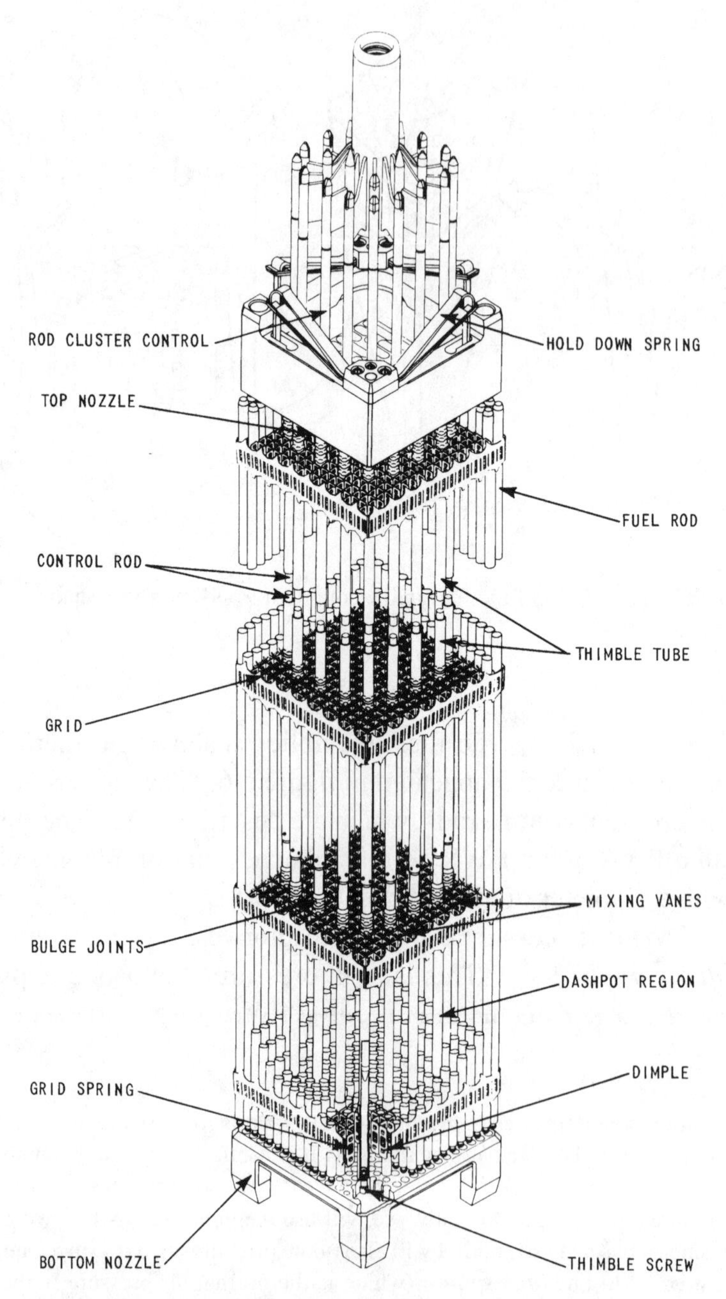

Figure 5.4 17 × 17 PWR fuel assembly. COURTESY OF WESTINGHOUSE ELECTRIC CORPORATION.

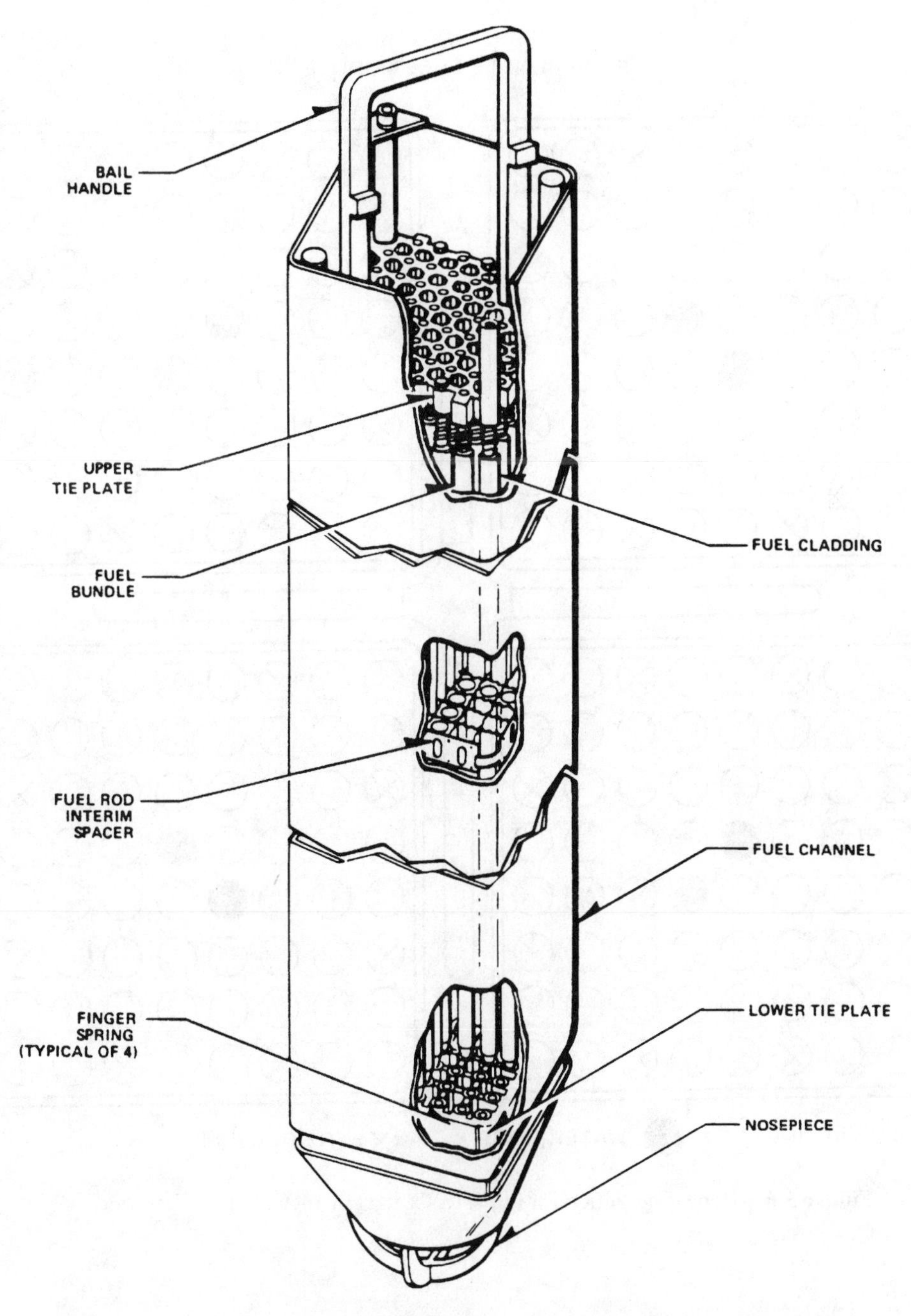

Figure 5.5 BWR fuel assembly. COURTESY OF GENERAL ELECTRIC COMPANY.

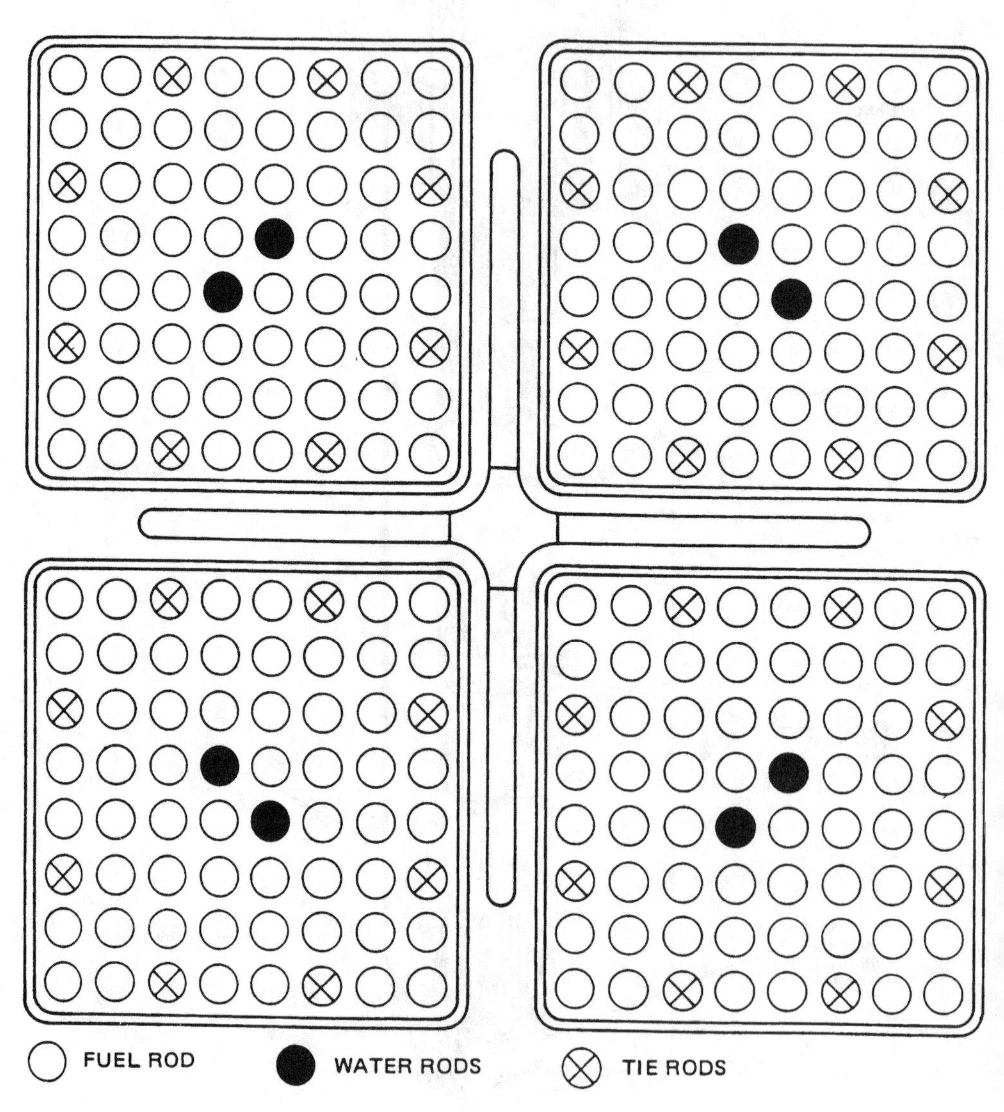

Figure 5.6 Repeating lattice for a BWR. COURTESY OF GENERAL ELECTRIC COMPANY.

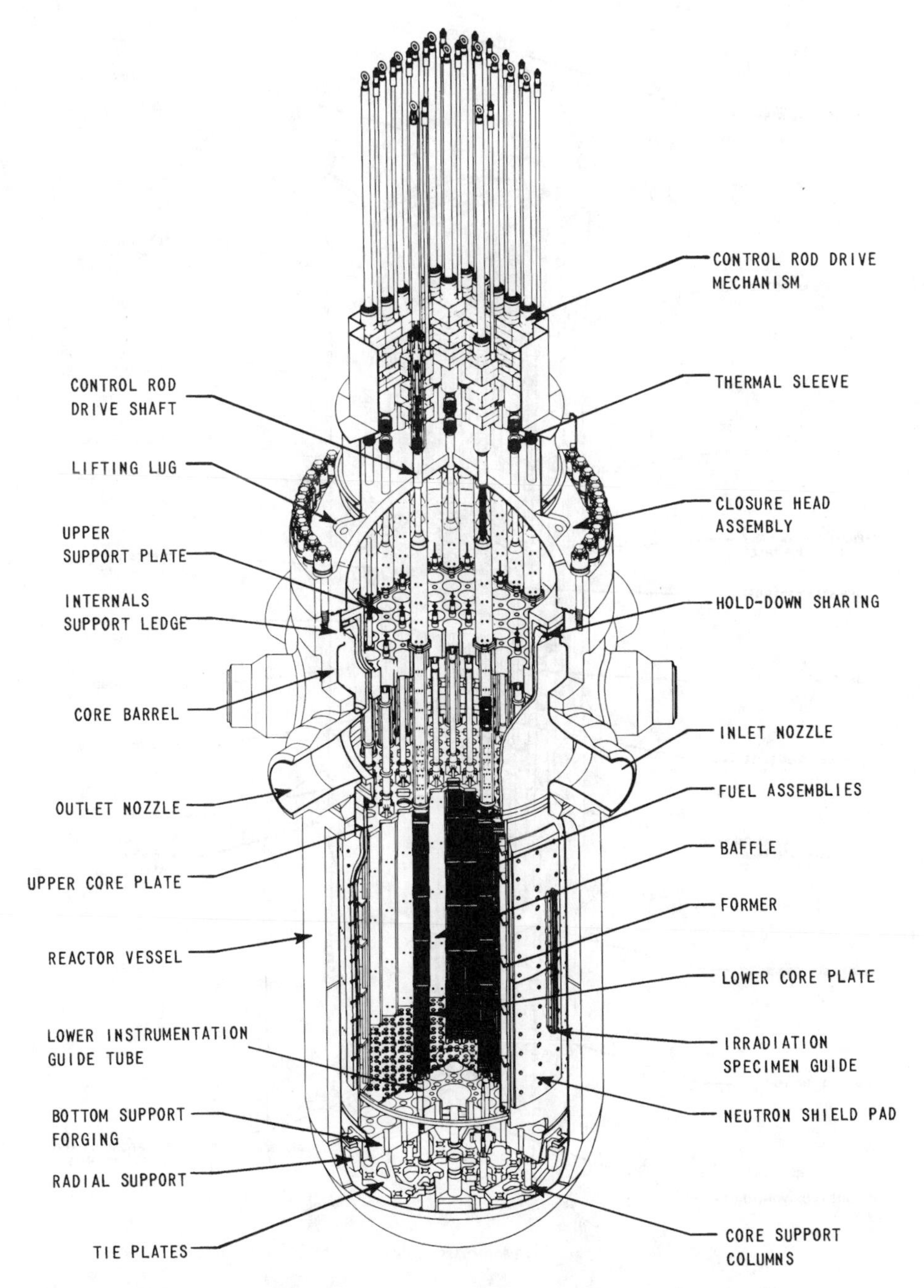

Figure 5.7 PWR core and reactor vessel. Courtesy of Westinghouse Electric Corporation.

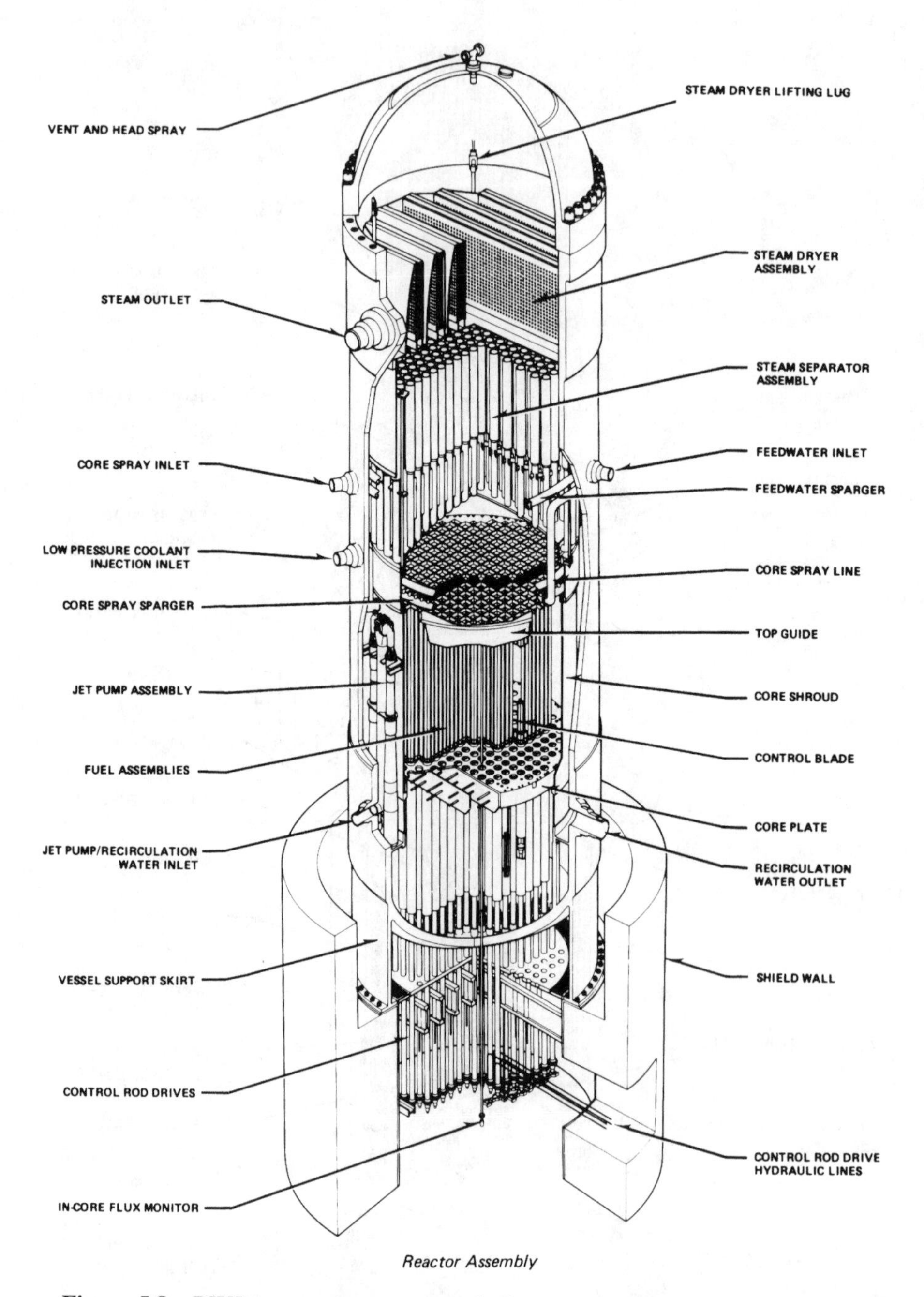

Figure 5.8 BWR core and reactor vessel. COURTESY OF GENERAL ELECTRIC COMPANY.

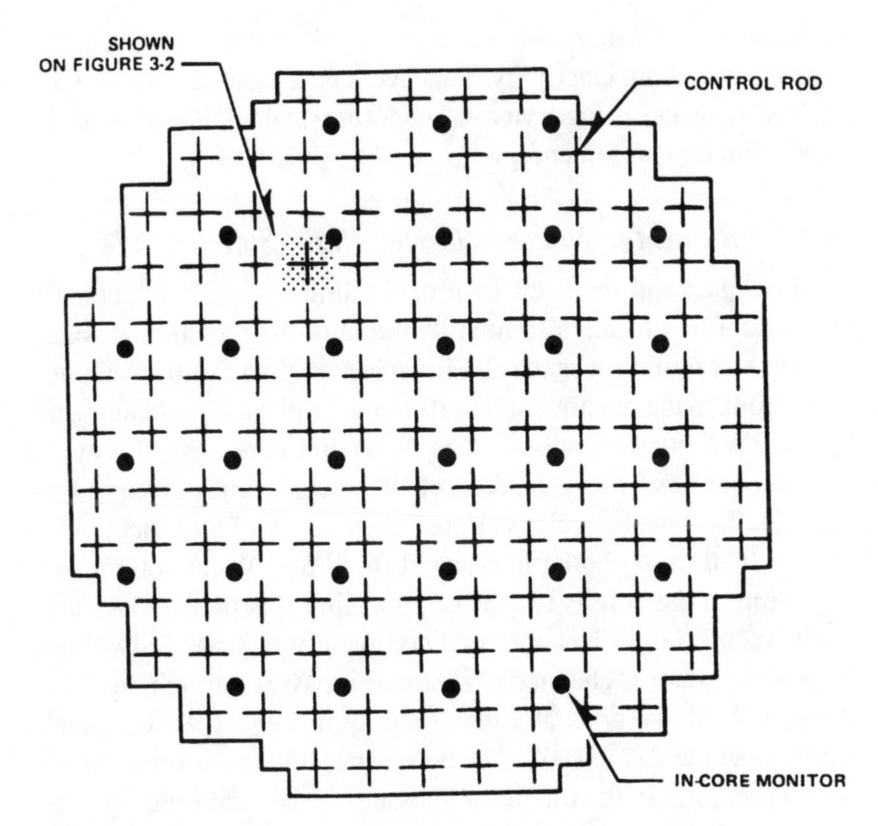

Figure 5.9 BWR core arrangement, top view. Courtesy of General Electric Company.

In metric units this is 15.1 megapascals, or 151 bars. The pressure in a BWR is also high, though not quite as high; it is 1050 psi (7.0 MPa, or 70 bars). Why such high pressures? It's a consequence of the steam cycle used to make the electricity, which is explained below. The PWR pressure vessel is about 40 ft (12 m) tall with an outer diameter of 14 ft and steel walls 8 inches thick. The dimensions of the pressure vessel of a BWR of comparable power are about 70 ft tall, 21 ft diameter, and 6 inch thick walls.

The water in a reactor serves several purposes. First, it is the ***coolant***, like the coolant in your automobile; it carries away the heat energy generated by fission from the high-temperature fuel. Second, the water has a basic role in the ***steam cycle***, which is the process needed for making electricity. This cycle is discussed next. A third role for the water is to serve as a ***moderator***, which is discussed in the section on neutron energy.

The BWR and the Nuclear Power Plant Steam Cycle

The flow path and steam cycle of the boiling water reactor, at least in its simplified form, is simpler than that of the pressurized water reactor, so I will discuss the BWR cycle first. In a BWR, water actually boils in the reactor core, as its name implies. The steam cycle for the BWR in its simplest form is shown in figure 5.10. This cycle is called the ***Rankine cycle***. A more detailed sketch is given in figure 5.11. This steam cycle is almost identical to a fossil fuel power plant, which uses a boiler in place of the BWR. Water coming out of the top of the core is two phase, meaning it is partly water and partly steam (i.e., it has not been completely boiled). Only about 14% of the water is changed to steam as it passes through the core. Water at 1050 psi boils at a temperature of 550°F (288°C), much higher than the 212°F (100°C) where water boils at normal atmospheric pressure. In the top of the pressure vessel, the water and the steam are separated in a device called, would you believe, a ***steam separator***. The steam is then dried in a ***steam dryer*** to remove any remaining moisture. These devices were shown in figure 5.8. The dry steam leaves the reactor vessel and becomes a part of the steam cycle, as shown in figures 5.10 and 5.11. The hot high-pressure steam passes through the turbine, where it impinges on the turbine blades. This causes both blades and turbine shaft to turn, like blowing on a pinwheel. This shaft is connected to the shaft of a generator that rotates coils of wire in a magnetic field to produce the electricity, which is the end product of the electric power plant. Notice from figure 5.11 that there is generally a high-pressure and a low-pressure turbine connected to the same shaft.

Now let's get back to the question of why there is such high pressure in a BWR. Why not just boil the water at atmospheric pressure and 212°F? The answer is that steam at a much higher temperature can generate electricity in the Rankine cycle far more efficiently

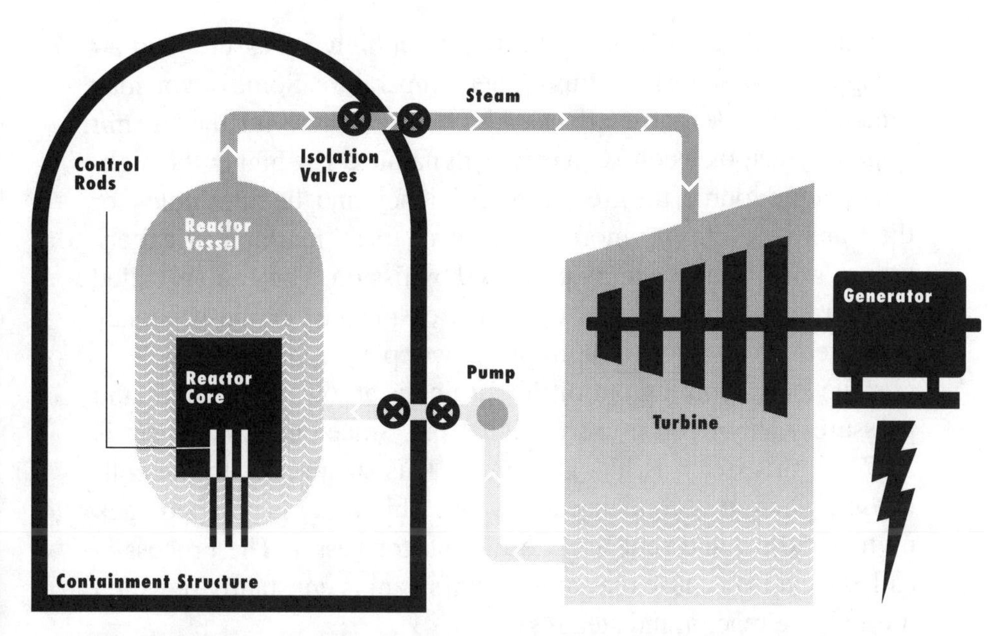

Figure 5.10 Simplified BWR steam cycle. COURTESY OF NUCLEAR ENERGY INSTITUTE.

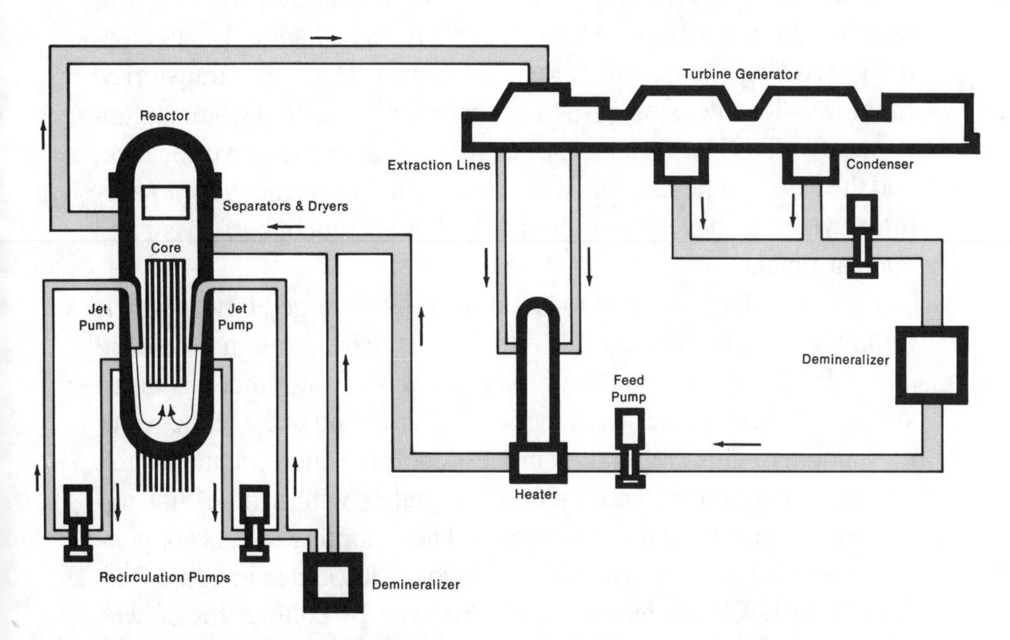

Figure 5.11 More detailed BWR steam cycle. COURTESY OF GENERAL ELECTRIC COMPANY.

than at 212°F. It is necessary to use much higher pressure than atmospheric to boil water at this higher temperature. Some of you may remember the desirable effect of high temperature for the ***Carnot cycle,*** if you took a course in thermodynamics. The higher the high-temperature source, the greater the efficiency, and the same holds for the Rankine cycle. By more efficiency, I mean more electrical energy per unit of heat energy produced by fission. This ratio is called the ***thermal efficiency*** of the cycle. Increasing the thermal efficiency is of great importance to running a power plant economically.

The steam coming out of the turbine is at low temperature and pressure—about 100°F (38°C) and 1 psi. Since one atmosphere is 14.7 psi, this steam is like a vacuum. This steam goes to the condenser, where it is condensed to water. This water is pumped back up to 1050 psi and sent back to the reactor vessel. The process is called a cycle because the same water/steam is constantly recycled through the reactor and steam system.

In the condenser, a large amount of heat must be removed from the steam in order to condense it. This heat is dumped into the environment. Yes, just dumped, to serve no useful purpose at all. In fact, it is a standard design problem that must be handled at all electric generating plants, both fossil and nuclear. The heat is transferred first to a cold water stream in the condenser. If the cold water stream comes from a river or lake (or ocean), it heats up the river or lake, and the extent of heating must be controlled. This is the ***thermal pollution*** you may have heard about, which is present in both fossil and nuclear plants.

The heat does not have to be dumped into a large body of water; it may instead be transferred to the atmosphere through a ***cooling tower.*** In this system, air flows past the warm water that has taken away the heat from the steam in the condenser and some of the water is evaporated. This evaporation process cools the unevaporated water, which is returned to the condenser, together with a little makeup water to replace what has evaporated. These cooling towers are generally very tall and concave in shape. Two are illustrated in figure 5.12 for a twin BWR nuclear station. This type of cooling tower was made famous in the accident at the Three Mile Island nuclear power plant since this was the method used there for dumping the waste

Figure 5.12 Nuclear power station in Pennsylvania, Limerick 1 & 2, with two boiling water reactors and natural draft cooling towers. COURTESY OF PECO ENERGY COMPANY.

heat, and the cooling towers were the most visible feature in the pictures of the plant. In fact, those pictures of Three Mile Island led many Americans to think of cooling towers as characteristic of nuclear plants. In reality these same cooling towers are used at fossil fuel plants just as often as at nuclear plants. The reason that the towers are so tall is that the air flow is by natural convection, or natural draft, just like the draft that occurs in a fireplace chimney. This tall a structure is needed to provide the required air flow. It is also possible to cool the condenser water by forced convection cooling towers, i.e., forced by electric fans. This is done at a PWR nuclear power station in Arizona where the availability of water is low, as shown in figure 5.13 with its three reactors. The electric operation of fans makes this method more expensive than others.

You might naturally wonder why the steam is condensed at all since it is going to be sent right back to the BWR to be boiled again. Well, there are a couple of reasons. The simplest is that we have to get the pressure back up to 1050 psi, and it is far cheaper to pump a liquid back up to that pressure than it is to compress a gas (such

Figure 5.13 Nuclear power station in Arizona, Palo Verde 1, 2, & 3, with three pressurized water reactors and forced convection cooling towers. COURTESY OF ARIZONA PUBLIC SERVICE COMPANY.

as steam). The second reason is more fundamental, though perhaps more difficult to grasp. In this cyclic process, if we had compressed the exhaust steam from the turbine back up to the pressure in the BWR without first removing heat from the steam, the steam temperature would be way too high to be used to cool the BWR fuel. Thus, it is absolutely necessary to remove much of the energy from the fluid (i.e., the steam/water) in the form of heat during each cycle in addition to converting some of the energy into useful work in the turbine. We wish we could convert all of the heat energy that we get from the nuclear reactor (or from a coal-fired boiler) directly into work and electricity at the turbine-generator, but nature (and the "second law of thermodynamics") won't allow us to do that. It's not that we engineers aren't very smart. We simply can't, no matter how smart we are. In fact, in nuclear and coal-fired plants we have to dump almost two-thirds of the heat energy to the environment, which means that we use only 33% to 40% of the energy in the turbine/generator. 33% is the thermal efficiency of a light water reactor plant; 40% is the efficiency of a modern coal plant. Natural gas "combined cycle" plants being built in the 1990s reach efficiencies approaching the 60% range. The gas-cooled nuclear plant, GT-MHR, described in chapter 7, is designed for 47% efficiency.

You will note in figure 5.11 a series of heaters between the condenser and the reactor. These ***feedwater heaters*** are used to increase the thermal efficiency of the steam cycle. This is accomplished by extracting some of the hot steam from the turbines and using it to heat the feedwater before it is returned to the reactor.

Recirculation in the BWR Pressure Vessel

I said that about 14% of the water in the BWR is converted to steam, the steam and water are separated, and the steam is sent to the turbine. Well, what happens to the 86% of the water that is not vaporized?

This water is circulated back to the bottom of the core. However, before it enters the core again, the 14% that was converted to steam and sent to the turbine is condensed. It comes back to the reactor as liquid water, called ***feedwater***, and is added to the 86% recirculated water. Thus 100% is again sent back through the core.

There is another complication, however. The recirculated water and feedwater that have joined together are at too low a pressure to pass through the core, due to friction and a resulting "pressure drop" as the water/steam mixture passes through the core. Thus the recirculated water must be sent to ***recirculation pumps,*** which raise the pressure of the water to the level necessary to flow through the core. The recirculation system is shown in greater detail in figure 5.14, together with a perspective view in figure 5.15. Also shown in these figures are ***jet pumps,*** which assist in increasing the pressure of the recirculating water.

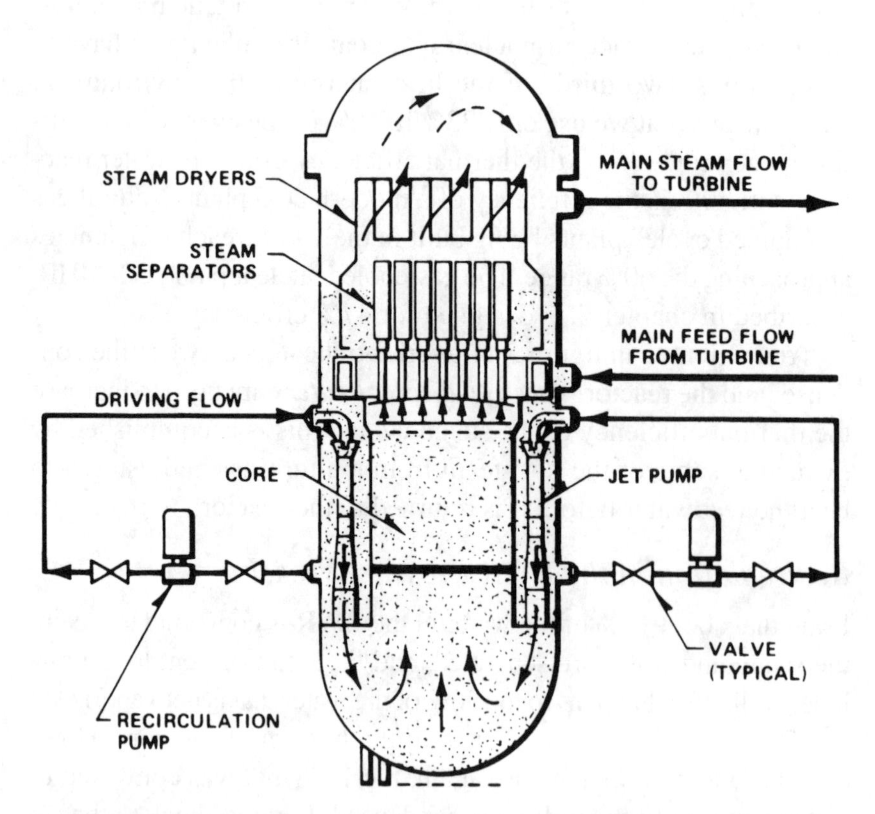

Figure 5.14 BWR recirculation system. COURTESY OF GENERAL ELECTRIC COMPANY.

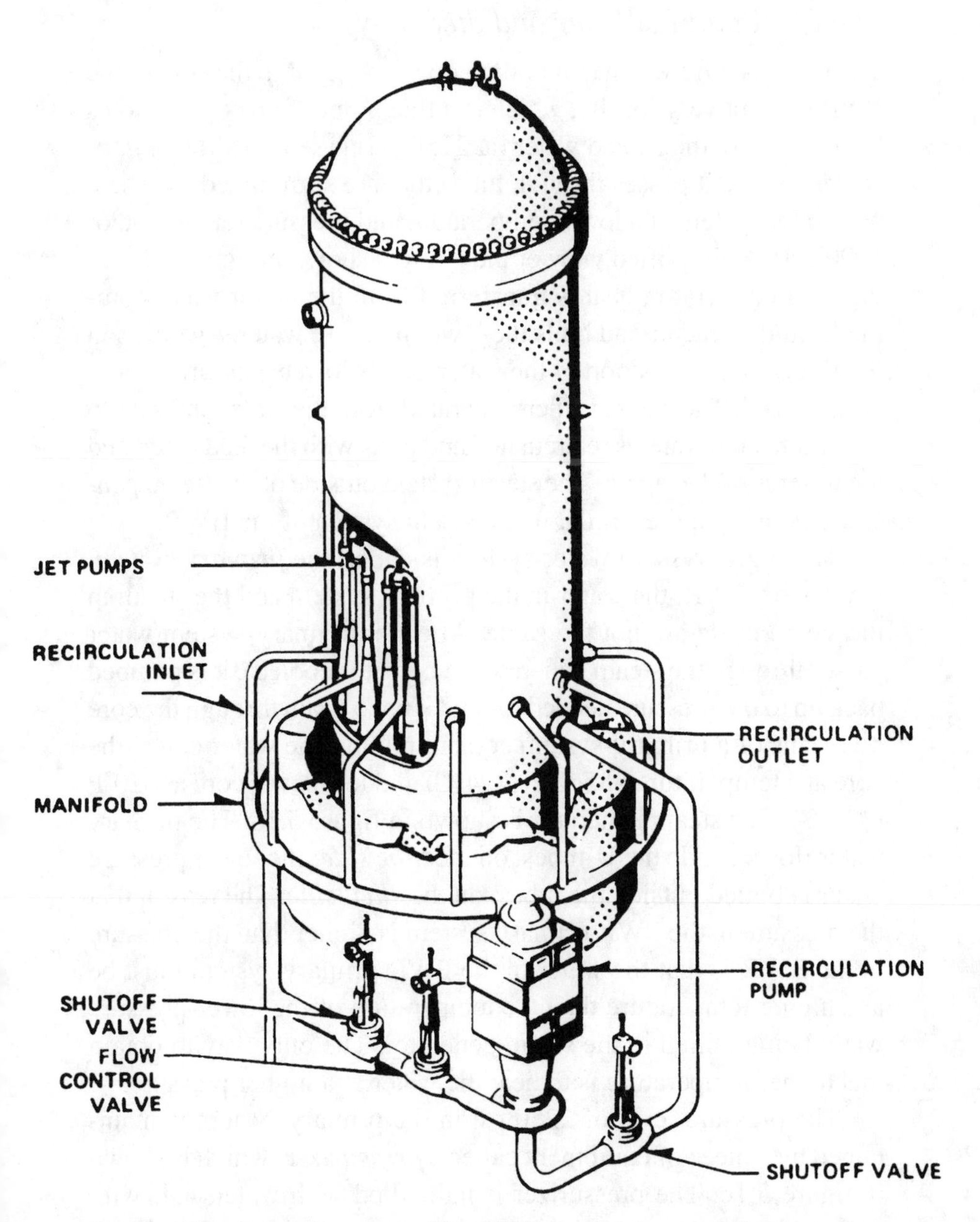

Figure 5.15 Perspective view of the BWR vessel and recirculation system. Courtesy of General Electric Company.

The PWR and Its Water and Steam Systems

The PWR is different in that boiling does not occur in the core of the PWR (except very locally, which is a fine point of no concern to us here). Instead, the heated water (at 2250 psi) is sent to a ***steam generator***, where it passes through tubes that are surrounded by a second water system at a lower temperature and pressure (about 900 or 1000 psi). A simplified view of the PWR is shown in figure 5.16. In the steam generator, heat is transferred from the hotter water coming from the reactor and boils the lower pressure water. Again, as in the BWR, only a fraction of the water in this lower pressure system is vaporized. The steam is then separated from the water and sent to the turbine; the water is recirculated and joins with the feedwater, and the water is boiled again. The steam system outside of the steam generator is really quite similar to the steam system of the BWR.

The high-pressure water system is called the ***primary system***. Unlike the BWR, the water in the primary system and the steam in the steam cycle are not the same. After the primary-system water passes through the steam generator (where it is cooled), it is pumped back up to the pressure needed to send it once again through the core by means of a primary-system coolant pump. The water enters the core at a temperature of 560°F (294°C) and leaves the core at 620°F (327°C). The steam generator is shown in figure 5.18. The primary water flows inside the U-tubes, on the ***tube side***; the lower pressure water is boiled outside the tubes, on the ***shell side***. The reason that the pressure in the PWR primary system is higher than the pressure in the BWR is that the water in the PWR primary system must be at a higher temperature than the temperature of the lower-pressure water being boiled in the steam generator. The only way to obtain this higher temperature is to heat the water at a higher pressure.

The pressure level of 2250 psi in the primary system is maintained by a piece of equipment called a ***pressurizer***, which is shown in figure 5.16. The pressurizer is half-filled with water, half with steam. If the pressure in the primary system needs to be increased, the water in the pressurizer is heated, which has the effect of raising both its temperature and pressure. If the pressure needs to be

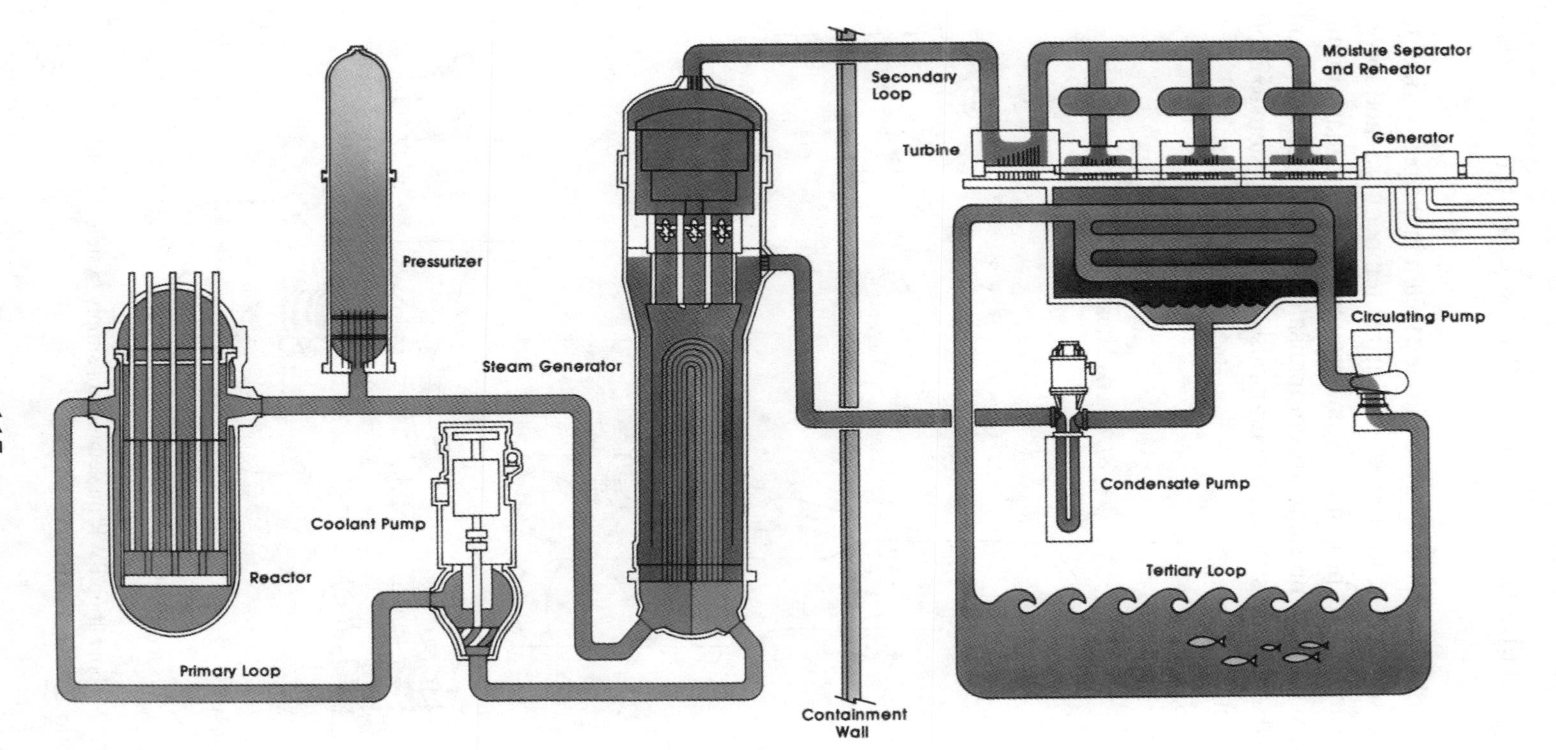

Figure 5.16 PWR system, with both primary system and simplified steam cycle. COURTESY OF WESTINGHOUSE ELECTRIC CORPORATION.

reduced, colder water is sprayed into the steam in the pressurizer to condense some of it, lower the pressurizer temperature, and lower the pressurizer and primary system pressure. (The primary pump adds only enough pressure to compensate for the pressure drop through the primary system; it cannot raise the whole system to 2250 psi by itself.)

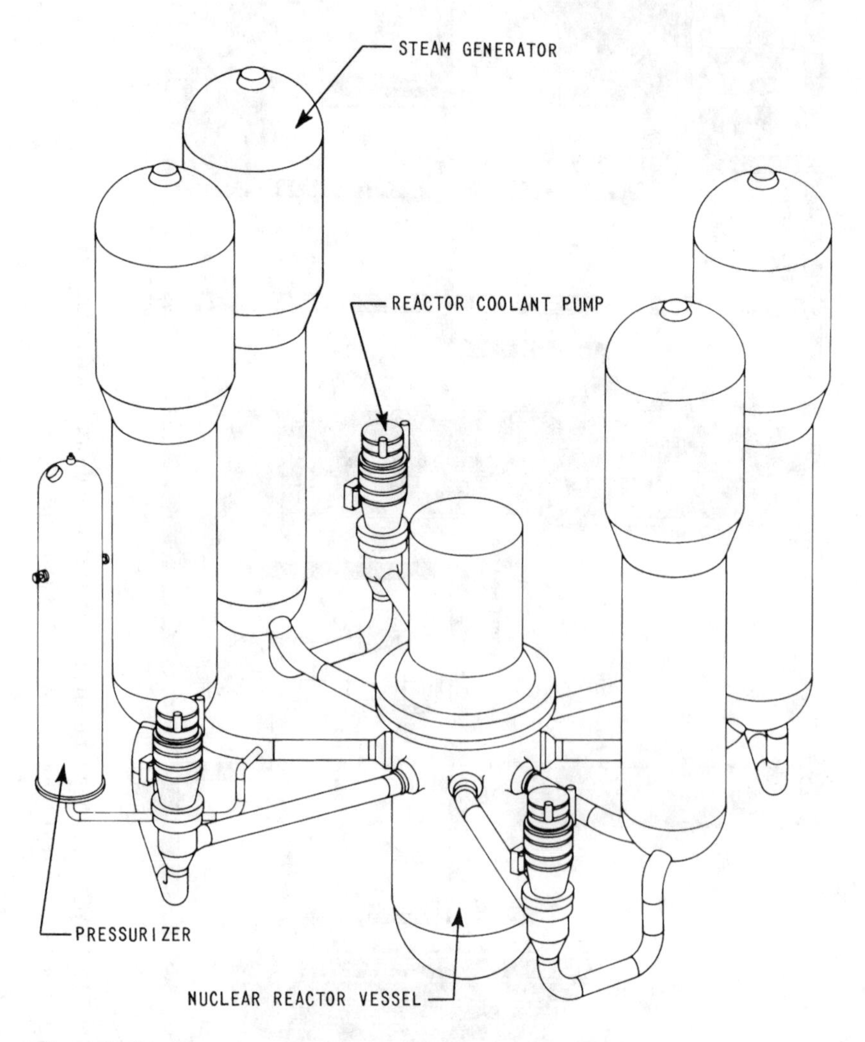

Figure 5.17 Four-loop PWR nuclear steam supply system. COURTESY OF WESTINGHOUSE ELECTRIC CORPORATION.

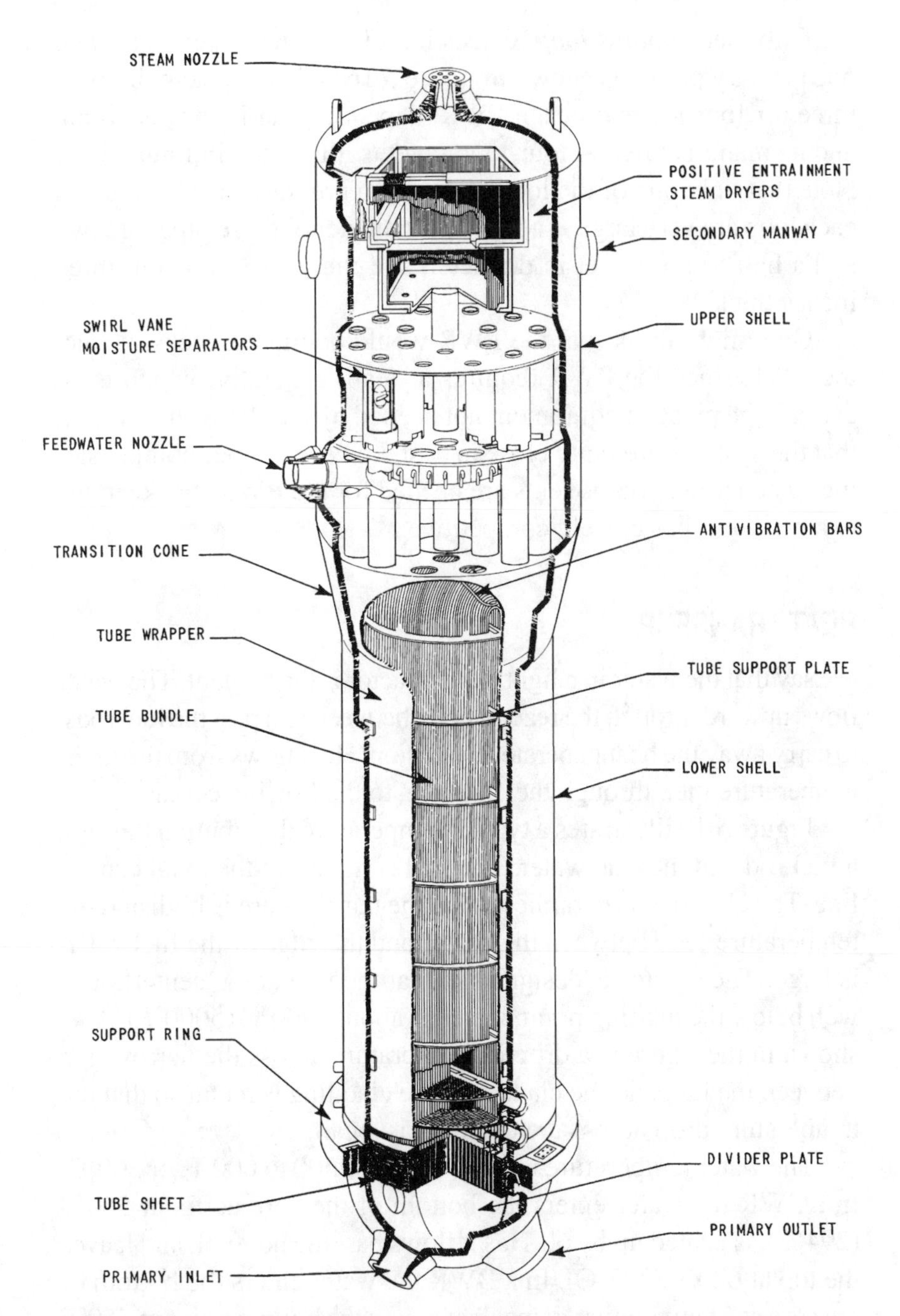

Figure 5.18 PWR steam generator. Courtesy of Westinghouse Electric Corporation.

Only one ***coolant loop***, consisting of one steam generator and one primary pump, is shown in figure 5.16. There are actually two, three, or four loops on each PWR, depending on its power rating and its manufacturer. A four-loop plant is illustrated in figure 5.17. Note that only one of the loops contains a pressurizer. The pipes in each loop are very large. The outside diameter of these pipes are two and a half to three feet in diameter, and the walls are about three inches thick.

One might think that the PWR would be more expensive than the BWR since the PWR requires a steam generator, which is an extra, large piece of equipment not needed by the BWR. It turns out that the systems are quite close in cost, however. Other compensating systems and components are needed for the BWR that keep the two reactor types closely competitive.

HEAT TRANSFER

We say that the water in a light water reactor is the coolant. The water flows upward through the reactor core, between and past the fuel rods, to carry away the heat generated by fission. Heat flows from the high-temperature fuel, through the cladding, to the flowing coolant.

Figure 5.19 illustrates a typical temperature distribution through a fuel rod and into the water. The fuel is hottest at the axial centerline. The UO_2 fuel is ceramic so that the temperature is high and the temperature drop between the center and the edge of the fuel pellet is large. The reactor is designed so that the fuel at the centerline is well below the melting point, which is about 2800°C (5000°F). Also shown in the figure is a drop in temperature across the narrow gap between the fuel and the cladding. The cladding is metal so that the temperature drop across the cladding is small.

The water temperature is in the range of 500 to 600°F, or 300°C. In a PWR the water enters the bottom of the core at about 560°F (294°C), is heated up by 60°F as it flows past the hot fuel, and leaves the top at 620°F (327°C). In a BWR the water enters the bottom of the core at a temperature somewhat below the boiling point of 550°F (288°C) at the system pressure of 1050 psi. The water is then heated

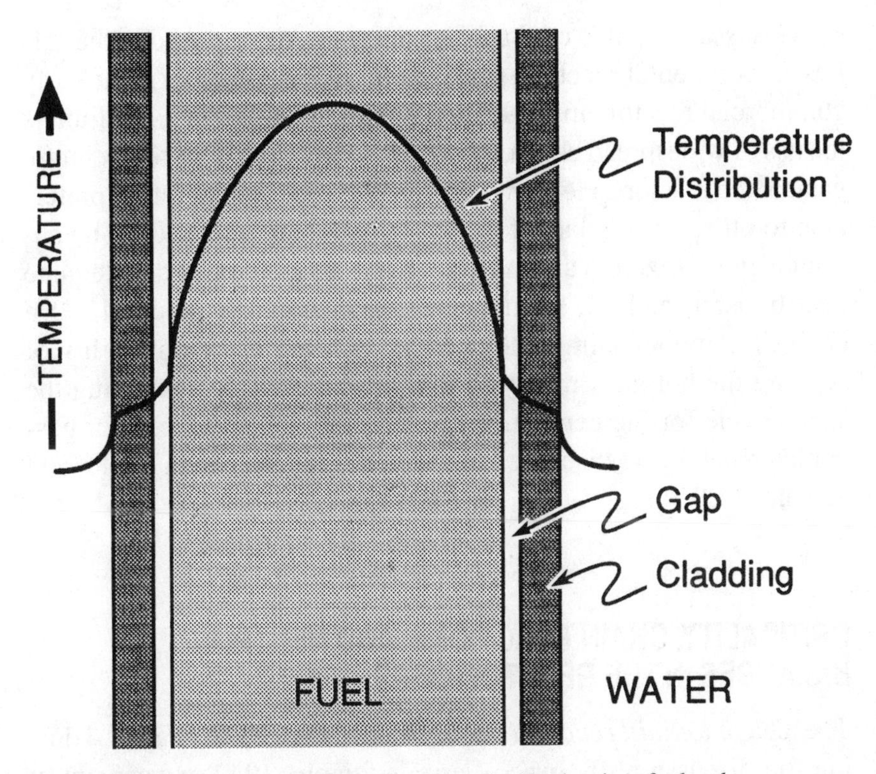

Figure 5.19 Representative temperature distribution in a fuel rod.

up to the boiling point in the lower part of the core, and then the temperature remains at the boiling point until it exits the core.

Other Coolants—Liquid Metal and Gas

The coolant for fast reactors (a type of reactor discussed later in this chapter and in chapter 6) is a liquid metal, sodium, which becomes liquid (melts) at 98°C (208°F). Thus the fast reactor is called the ***liquid metal reactor*** (LMR). (As an interesting aside, this reactor used to be called the liquid metal fast breeder reactor in the days when breeding was stressed. Politically speaking, however, breeding is in not in vogue at this particular time, so we engineers changed the name. You see, we are not immune to the political winds any more than the rest of society when funding of programs is concerned. Breeding is discussed at the end of this chapter.)

Gas is also used to cool nuclear reactors. Air was used in the earliest experimental reactors. Carbon dioxide was used in some early commercial reactors in the United Kingdom and France. Helium is the gas of widespread choice, however, for modern high-temperature gas-cooled reactors. Helium is chemically inert and therefore preferable to other gas coolants for high-temperature operation. The helium is pressurized to around 1000 psi in order to provide adequate heat transfer, and the helium outlet temperature is about 1300°F (700°C). An interesting feature of gas cooling is the possibility of sending the hot gas directly to a gas turbine rather than using the steam cycle for the generation of electricity. New designs for gas-cooled reactors are discussed in chapter 7, together with LMRs and advanced LWRs.

CRITICALITY, CHAIN REACTIONS, AND NEUTRON BALANCES IN THE REACTOR CORE

The idea of a ***chain reaction*** was introduced briefly in chapter 4 during the discussion of Enrico Fermi's famous 1942 experiment at Stagg Field. In a chain reaction, one event leads to another, then another, and so on. A neutron causes fission, and the fission produces more neutrons, and these neutrons cause more fissions, ad infinitum. This is a fission chain reaction. The term is applied to many things nowadays, like an automobile pileup in the fog on the interstate.

One can conveniently think of generations of neutrons. One generation of neutrons causes fissions, and the neutrons produced thereby become part of the next generation, which produces fissions in the next generation. An objective in the operation of a nuclear reactor is to maintain a constant power level (i.e., the rating in megawatts). This means keeping the number of fissions in each successive generation constant. The power level is proportional to the rate at which fissions occur, so keeping the number of fissions in each generation constant keeps the power level constant. If the number of fissions in each generation remains constant, the reactor is said to be ***critical***. Recall that one neutron causes a fission, but between 2 and 3 neutrons are produced per fission. Therefore, at least one of these neutrons must cause a fission in the next generation in order for the chain reaction to proceed, and exactly one must cause fission for the power level to remain constant and the reactor to be critical.

The question then arises, what happens to the other neutrons? Two things. First, neutrons can be captured by other nuclei and not cause fission. Second, neutrons can leak from the reactor core.

These events, together with fission, can be conveniently expressed mathematically as a ***neutron balance***. In a critical reactor, the neutron population must neither rise nor fall. The critical neutron balance equation states that the rate of production of neutrons by fission must be exactly equal to the rate of loss of neutrons by leakage and absorption. In words, this equation is:

$$\text{Neutron leakage rate} + \text{Neutron absorption rate} = \text{Neutron production rate.}$$

To use this equation, we need expressions for the rates of leakage, absorption, and production. In a reactor core, a nuclear reaction rate is the rate (per unit volume in the reactor core) of reaction between a neutron and a nucleus, like uranium, expressed as follows:

$$\text{Reaction rate} = N\sigma\phi \ (= \text{nuclear reactions/cm}^3 \cdot \text{s}) = \Sigma\phi$$

where

N = atoms of nuclei per cm^3 of reactor core volume.
ϕ = neutron flux (neutrons/$\text{cm}^2 \cdot \text{s}$), a quantity proportional to the density of neutrons.
σ = microscopic cross section, a property of the nucleus proportional to the probability of the occurrence of the reaction, in units of cm^2, or more often, reported in the unit of ***barns***, where 1 barn = 10^{-24} cm^2.
Σ = macroscopic cross section (in units of 1/cm), and $\Sigma = N\sigma$.

The reaction rates in the neutron balance equation are:

$$\text{Leakage rate} = DB^2\phi$$

where

D = diffusion coefficient (cm), which is related to the distance a neutron travels before reacting with some material in the reactor.
B^2 = buckling (cm^2), which is related to the physical size, or geometry, of the reactor.

$$\text{Absorption rate} = \Sigma_a\phi$$

where Σ_a = absorption cross section, which includes both fission and non-fission capture of neutrons.

$$\text{Production rate} = \nu\Sigma_f\phi,$$

where

Σ_f = fission cross section (so that $\Sigma_f\phi$ is the fission rate).
ν = number of neutrons produced per fission ($\nu = 2.44$ for U-235).

Therefore, the neutron balance equation for a critical reactor is

$$DB^2\phi + \Sigma_a\phi = \nu\Sigma_f\phi$$

Let me give you some feeling for the magnitudes of these quantities in a typical pressurized water reactor. The average thermal neutron flux, ϕ, is about 3×10^{13} n/cm$^2 \cdot$ s. Most of the fissions occur at thermal energy so this is the energy that I am stressing here. I say "average" flux because the flux actually varies with position in the core. The m*a*croscopic thermal fission cross section, Σ_f, is about 0.09 cm^{-1}, which is equivalent to a U-235 atom density of 2×10^{20}

atoms/cm^3 (and no fissile plutonium) and a U-235 *mi*croscopic thermal fission cross section, σ_f, of 450 barns (or 450×10^{-24} cm^2). The product $\Sigma_f\phi$ is 3×10^{12}, which is the number of fissions per second per cubic centimetre in the core. In a typical large PWR or BWR, about 40% of the neutrons cause fission, nearly 60% are captured in non-fission events, and only 2% to 4% leak from the reactor core.

The neutron balance equation suggests the introduction of a new parameter called the ***criticality factor***, or ***multiplication factor***. The criticality factor, k, is obtained by dividing the neutron production rate by the neutron loss rate, or

$$k = \frac{\nu\Sigma_f}{DB^2 + \Sigma_a} = \frac{\textit{neutron production rate}}{\textit{neutron loss rate}}$$

If k = 1, the reactor is critical. If k < 1, the reactor is subcritical. For this case the neutron production rate is less than the loss rate and the neutron level will fall. If k > 1, the reactor is supercritical. The production rate is greater than the loss rate, and the neutron level will rise.

A related useful concept is ***reactivity***. Reactivity, ρ, is a measure of how much the criticality factor differs from unity. Hence,

$$\rho = \delta k/k = (k - 1)/k.$$

For a critical reactor, $\rho = 0$; for a subcritical reactor, ρ is negative; and for a supercritical reactor, ρ is positive.

You have heard the term ***critical mass***. This is the amount of fissile material, like U-235, that makes the reactor exactly critical, or k = 1. You can calculate the critical mass from the neutron balance equation or the equation for k given above. Suppose you have a particular size reactor; this fixes the buckling, B^2, since buckling is dependent only on the geometry. Then you calculate the D, Σ_a and Σ_f, varying the atom density of the fissile material until k = 1 (i.e., until the neutron balance is exactly satisfied). Multiplying this atom density times the reactor volume, together with conversion factors for atoms to kilograms, gives you the critical mass.

The uranium in a light water power reactor (LWR), in both the PWR and the BWR, is typically enriched in U-235 to about 3% to 4%—up from the 0.7% U-235 in natural uranium. This high a concentration of U-235 is needed in order to make the reactor critical. Power reactors have been built that use natural uranium, but they must use either heavy water or gas cooling and graphite, which is a form of carbon, in place of ordinary light water in order to decrease the nonfission capture sufficiently to achieve criticality with such a low U-235 content. (Normal hydrogen in water absorbs neutrons considerably more than heavy hydrogen or graphite. The purpose of the graphite is discussed below in the section on moderators.) Most of the western world's power reactors are light water reactors, however, because the overall economics of the LWR are more favorable, despite the added expense of enriching the fuel to 3%. The Canadians, who make an excellent and economical heavy water reactor that uses natural uranium, would take issue with this assessment, however. Indeed they have sold some of their CANDU reactors to countries outside Canada. You see, things are rarely clear cut, even in the realm of science and engineering.

An especially interesting reaction in the reactor core is the capture of a neutron by uranium-238. This abundant isotope of uranium is useless by itself. However, when it captures a neutron, it becomes U-239. Uranium-239 is radioactive; it decays rapidly to neptunium-239 by ejecting a beta particle. The Np-239 then decays further by

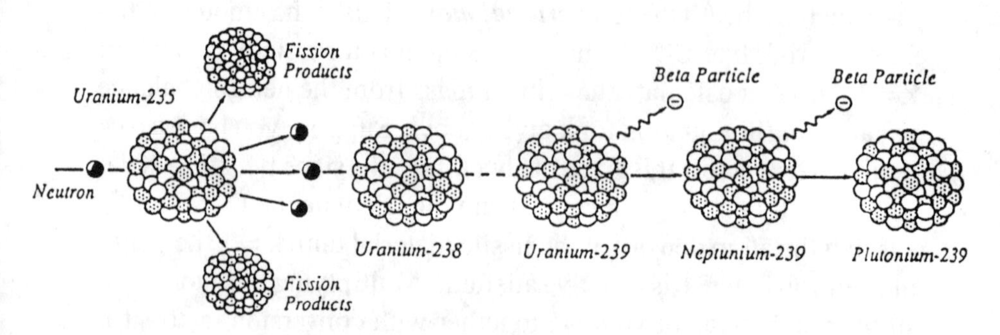

Figure 5.20 The formation of plutonium. COURTESY OF DOE.

ejecting still another beta to become plutonium-239. This process is illustrated in figure 5.20. Now Pu-239 is a very valuable fuel. Like U-235, Pu-239 can be fissioned by thermal neutrons, so it is a useful fuel in a thermal reactor, such as an LWR. A large amount of Pu-239 is produced in an LWR so that about 40% of the fissions in an LWR actually occur in Pu-239. In fuel that is removed from an LWR after its useful life, the Pu-239 content is about the same as the U-235 that still remains in the fuel, with each accounting for about 0.5% of the fuel.

NEUTRON ENERGY AND MODERATORS

Thus far we have generally ignored the influence of neutron energy in our description of the nuclear reactor.

As mentioned in chapter 4, neutrons that are produced during fission range in energy from a few tenths of an MeV to 14 MeV. The average energy of these fission neutrons is about 2 MeV.

The thermal neutrons that cause most of the fissions in a thermal reactor are in the energy range around 0.02 eV. Thus there is a factor of 10^8 difference between the energy of fission neutrons and thermal neutrons: 2 MeV/0.02 eV = 10^8. The question therefore arises, how do the high-energy neutrons become thermalized?

Finally we are coming to the subject of ***moderators***. The process of reducing the energy of the neutrons is called ***moderation***. When neutrons strike the nucleus of an atom, the neutron is not necessarily captured by the nucleus. It is more likely that the neutron will bounce off the nucleus; we say that the neutron is ***scattered*** by the nucleus. If the neutron bounces off like two billiard balls colliding in which no energy is lost in the collision, the event is called ***elastic scattering***. Both kinetic energy and momentum are conserved in such a collision. If the target nucleus is small, the neutron can lose much of its energy in a single elastic collision. You can imagine this. If a billiard ball hits a second billiard ball head on, the first ball will stop dead in its track because the balls are of the same mass; it will impart all of its energy to the second ball. Similarly, if a neutron collides elastically head-on with a hydrogen nucleus, it will lose all of its energy in this single collision. Of course, it probably will not hit head-on, which means that it will lose only a fraction of its energy. Thus a number of collisions will be required to reduce the energy of a high-energy neutron all the way down to thermal energy. On the other hand, if a neutron suffers an elastic collision with a large nucleus, it cannot lose much energy even if the collision is head-on. This is like a billiard ball hitting a bowling ball. Since we want to thermalize the neutrons in a thermal reactor, we want to have many light nuclei available for the neutrons to strike in order to thermalize them with as few collisions as possible. The material that we add to a reactor for this purpose is called the ***moderator***.

Since light elements can reduce the neutron energy faster than heavy elements in elastic scattering collisions, moderators are always selected from the light elements. The main candidates are normal hydrogen, deuterium, and carbon (in the form of graphite). Beryllium has also been used. Hydrogen and deuterium are available in the form of water—H_2O and D_2O. The most important scattering collision by a neutron in H_2O is a collision is between the neutron and the hydrogen nucleus, i.e., between particles of nearly equal size. The oxygen in water provides some moderation, but not much compared to the hydrogen. In a gas-cooled reactor the preferred moderator is graphite.

Both graphite and deuterium have the attractive feature of having very low capture cross sections, and hence low propensity for neutron capture, even after the neutrons have been thermalized. Normal hydrogen has a larger capture cross section, so it competes with the uranium for the thermal neutrons. On the other hand, light water is inexpensive and can moderate neutrons in the smallest volume. Water, both light and heavy, can fill the dual roles of both coolant and moderator; in a gas-cooled reactor, helium is generally the coolant of choice, and graphite serves only as a moderator. Since carbon is more massive than hydrogen, and consequently more elastic collisions are required for moderation, the space required for carbon to moderate neutrons is much larger. Therefore, gas-cooled reactors are much larger in size than water-moderated reactors.

Our discussion so far has concerned elastic scattering collisions. Sometimes, for high-energy neutrons, the collision is accompanied by the emission of a gamma ray from the target nucleus. In this case the kinetic energy of the neutron and the target nucleus are not conserved, and the neutron can lose a significant fraction of its energy even in a collision with a heavy nucleus. This event is called ***inelastic scattering***. It is also important in nuclear reactors, though not so important as elastic scattering.

Thus far we have restricted the discussion to thermal reactors. A second type of reactor is a ***fast reactor***. In this type of reactor, the fissions are caused by high-energy, or fast, neutrons. Thus no moderator is present in a fast reactor; one does not want the neutrons to slow down. They actually do slow down to some extent due to inelastic scattering and some elastic scattering with medium-size nuclei. Fast reactors are currently not as economical as thermal reactors, so they are not being used commercially at present. More will be said about fast reactors in chapter 7.

CONTROL

The neutron population and, hence, the power level of the reactor, are controlled, or regulated, with ***control rods***. These are rods of a material that has a high propensity for capturing neutrons, and thus has a high capture cross section. Boron and cadmium are two such materials. These materials are appropriately called ***neutron poisons***. Boron carbide imbedded in steel cruciform blades is used in the control rods of a BWR. An example is shown in figure 5.21. These blades fit between the fuel assemblies. The control rods of a PWR are rods of a silver-indium-cadmium alloy that fit inside the fuel assemblies in place of the normal fuel rods. A control-rod assembly consists of around 16 to 24 of these rods, as shown in figure 5.22.

When the reactor is shut down and, hence, not operating, the control rods are inserted. The nonfission capture is so great that the reactor is subcritical, i.e., the criticality factor, k, is less than one. In order to bring the reactor up to power, the control rods are first

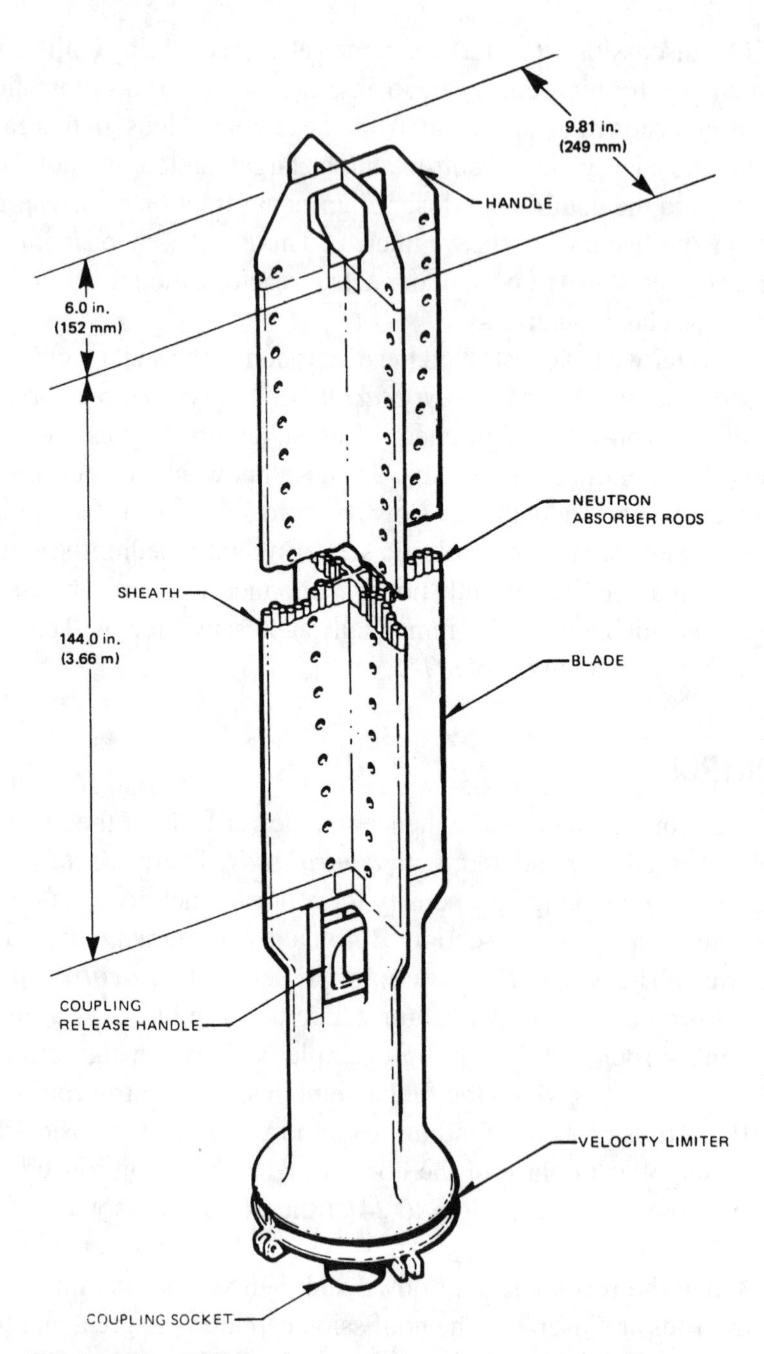

Figure 5.21 BWR control rod. COURTESY OF GENERAL ELECTRIC COMPANY.

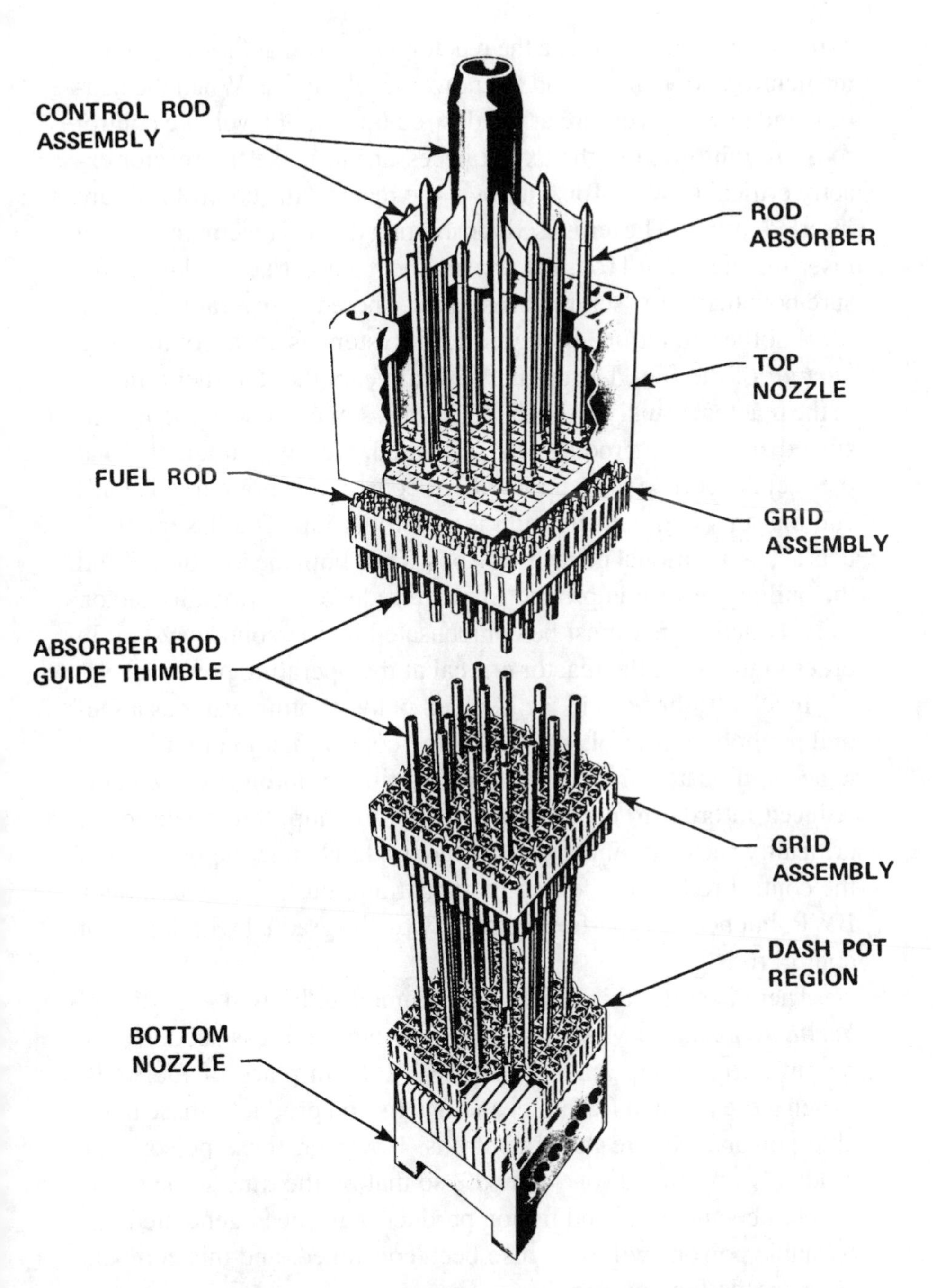

Figure 5.22 PWR control rod assembly. Courtesy of Westinghouse Electric Corporation.

withdrawn enough to make the reactor supercritical (k > 1) to allow the neutron population (and the power level) to rise. When the neutron and power levels are at the desired operating level, the control rods are reinserted to the extent necessary to make the reactor exactly critical (k = 1). To shut down the reactor, the control rods are inserted further. The term used for shutting down a reactor by rapidly inserting the control rods is to ***scram*** the reactor. (No one knows for sure how the term "scram" originated. There are several theories.)

Another function of the control system is to accommodate ***burnup*** of the fuel. During the several years that the fuel remains in the reactor (usually three to five years), some of the U-235 is consumed and fission products are generated. We say colloquially that the fuel is "burned" (even though, of course, it is not burned at all) and that this fuel consumption is called burnup. The fission products act as additional neutron poisons. Thus both the loss of fuel and the buildup of fission products tend to reduce the criticality factor. This reduction in k must be compensated by the control system in order to maintain the reactor critical at the operating power level.

In a PWR, boric acid is dissolved in the cooling water as a second method of control, and this is the control method that is used to accommodate burnup. The concentration of boron in the water is reduced in order to compensate for fuel burnup, thus keeping the criticality factor at unity without having to change the position of the control rods. Compensating for burnup must also be done in a BWR, but here it is performed mostly by the gradual withdrawal of control rods.

There is still a third way to accommodate burnup—the use of ***burnable poisons***. Materials with high capture cross sections can be put into a few permanently placed rods in place of fuel rods. When there is much fresh fuel and few fission products, these burnable poisons capture many neutrons. However, these poisons are gradually consumed (or "burned") so that by the time some of the fuel has been burned and fission products have been generated, the burnable poisons will have also been consumed, and this compensates for the fuel burnup.

Now for those who enjoyed the discussion a few pages back about neutron balances and criticality factors, there is another facet to control that you will find equally titillating. If you think about it, you might raise the question, how fast can the power of the reactor be raised if you withdraw a control rod? We want to be able to raise the power slowly so that we are sure to have effective control of the reactor at all times. After all, you can conjure up the extreme situation of a nuclear weapon where the power rises fantastically fast, which is the whole point of a weapon design. Obviously we can't have that. As our good fortune would have it, the world was made in such a remarkable way that this is quite easily avoided; ***in fact, it is absolutely impossible for a reactor to explode like a nuclear weapon***. (The reason for this fact is discussed in chapter 11.)

The reason that a reactor is easily controlled, meaning that it is easy to raise the power slowly and with stability, is because of ***delayed neutrons***. These are in contrast to ***prompt neutrons***. Most of the 2.44 neutrons produced in the fission of U-235 are produced immediately during the fission process. It takes only a fraction of a millisecond (about 0.02 ms in an LWR) for these neutrons to slow down and cause another generation of fissions. If all neutrons were prompt, it would be difficult indeed, in fact impossible, to control the rate of rise in power by mechanical means like control rods since everything would be happening in milliseconds. Here is where the delayed neutrons come in to save us. Delayed neutrons are produced seconds, some even tens of seconds, after the fission event. In the fission of U-235, only 0.64% of the neutrons are delayed. At criticality, 99.36% of the neutrons being produced are prompt, but the remaining tiny fraction of delayed neutrons is still needed to sustain the chain reaction. Remember, the criticality factor has to be exactly 1, not 0.9936, for the reactor to be critical. As long as we change the control rods so that the change in criticality is less than 0.64% (i.e., k is less than 1.0064), the rate of change of the power is controlled by this tiny fraction of delayed neutrons, which means that the power change occurs over seconds instead of milliseconds. This tiny fraction is all that is needed.

Now, is it easy to be sure that we change the criticality by only such a small percent when we pull a control rod? The answer is yes. Very definitely yes. No question about it. That's exactly how all reactors are controlled, and we've been doing it now for more than 50 years. And it's very easy. We always let the science teachers in our week-long summer course at the University of Virginia pull the control rods from our research reactor and watch the power slowly rise (under the watchful eye of a licensed operator, of course; I don't mean to imply that we are cavalier about these things). It's almost like magic that nature provided us with this tiny fraction of delayed neutrons, but that's the way it is. The delayed neutrons actually come from the decay of several fission products, so that they are somewhat indirectly made from fission.

Ah, but naturally you ask, ***what if*** you change the criticality ***more*** than the magic 0.64%? Then you go ***prompt critical***, which means that you are critical on prompt neutrons alone, meaning that you don't need the delayed neutrons any longer to sustain criticality. Then, indeed, the power would rise very rapidly, on a millisecond scale, too fast to control with mechanical things like control rods. Reactors are designed so that it is easy to operate without ever getting close to prompt critical. But it is not impossible. So then, you ask, if, say accidentally, you do exceed prompt critical, then what is to prevent the reactor from running away, or, heaven forbid, even acting like a bomb?

Again nature is kind to us. It is still impossible for the reactor to act like a bomb. There are inherent mechanisms in reactors that make this impossible, mechanisms not present in nuclear weapons. As the power level rises, the temperatures of the fuel and the coolant also rise. In all of the western commercial reactors, as these temperatures rise, several inherent (i.e., without operator action) mechanisms immediately operate to reduce the criticality factor down below prompt critical. This feature is called having a ***negative power coefficient***. The fastest acting mechanism, called the ***Doppler effect***, is an increase in the absorption of neutrons by U-238 in the fuel as the fuel temperature rises. (As you science types know, the Doppler effect causes the sound of train whistles to change due to frequency changes depending on the relative motion of the train, i.e., toward you or away from you. The nuclear effect operative here is called a Doppler effect because it depends on changes in the relative speed between neutrons and U-238 nuclei as the fuel temperature changes.) This added non-fission absorption in U-238 quickly throws the neutron balance below prompt critical, abruptly halting the rapid rise in power. (In a nuclear weapon there is no U-238—or insufficient U-238—to provide this function, but it is present in all commercial reactors.) Other mechanisms operating to change the neutron balances and, hence, to slow down the rate of power rise include fuel expansion, coolant temperature rise, and even coolant boiling.

Next, you might ask, is there any experimental evidence that all this really works? Have we made reactors more than prompt critical to show that they do in fact shut themselves down automatically? The answer to each question is yes. In the 1960s I personally worked on a fast reactor called SEFOR, built by General Electric in Arkansas, for which the primary purpose was to demonstrate that the reactor would shut itself down by means of the Doppler effect if its criticality factor were suddenly made greater than prompt critical. A boron rod was purposely blown out of the reactor to make the reactor exceed prompt critical. In test after test the reactor performed exactly as it was calculated to do. In addition to this personal experience, I can report that many thermal test reactors operate to provide either rapid bursts of neutrons or rapid heating of fuel for various research experiments by purposely exceeding prompt critical, only to be shut back down quickly and automatically without any insertion of control rods.

In those days, I was a young manager at General Electric in charge of the SEFOR physics design. I reported to Paul Greebler, one of the world's leading reactor physicists and one who developed much of the theory for calculating the Doppler effect in fast reactors. The engineer in charge of the SEFOR project was Bertram Wolfe, who later headed up all of the nuclear business at General Electric. SEFOR was actually a joint project between the U.S. and several organizations in Europe. The scientist in charge of the European group was Wolf Häfele, who later became head of one of Germany's national laboratories and for many years was a leading figure in fast reactor development in Europe. The first time I ever traveled to Europe was with the General Electric team to make presentations on SEFOR at a meeting at the Karlsruhe Nuclear Research Center in Germany. I will never forget the experience of listening to engineers from five different European countries all making presentations in English since that was the only language any of us from the United States knew.

FUEL MANAGEMENT

Having talked a few pages back about fuel burnup, I would like to introduce a new topic called ***fuel management***. This has to do with how long the fuel remains in a nuclear reactor, or how often the burned fuel must be replaced with new fuel, a process called ***refueling***. One of the remarkable advantages that a nuclear plant has over a coal plant is that refueling a nuclear plant occurs only once every year or two, while refueling of a coal boiler is constant. A large coal plant requires a trainload of new coal every day. A nuclear plant requires a few truckloads of fuel every year or two.

A fuel assembly will remain in the reactor for about three to four years. At the end of that time the U-235 content is low, the fission product inventory is high, and the zircaloy cladding has suffered radiation damage from the hostile neutron environment. The combination of these factors requires that the fuel be discharged and replaced with ***fresh fuel***. The fuel being discharged is called ***spent fuel***. Reactors are shut down for refueling once every year or eighteen months and about one-third of the fuel is replaced.

During refueling, the top (or head) is removed from the reactor vessel. The spent fuel is then removed from the reactor and transported to a ***spent fuel storage pool***, where it is stored until it is cool enough to send to dry storage. This fuel must be kept under water at all times during the transfer to the storage pool since the fission products are still producing heat and radiation. To keep the fuel under water during refueling, the ***reactor cavity***, which is the space around and above the reactor vessel, is temporarily filled with water. The large quantity of water needed to flood the reactor is kept in a ***refueling water storage tank*** (RWST), located outside the containment building in present reactors. More will be said about this water storage tank in chapter 7.

A goal in fuel manufacture and reactor operation is to make the fuel last for increasingly long periods of time, but fuel management decisions are governed by economics as well as materials. It is possible to make the fuel last for a very long time. In the case of nuclear submarines, the fuel is designed to last for roughly fifteen years, but economics is not an important factor in this decision. However, economics is all important in commercial reactor operation, so compromises are required between length of time between refueling and fuel enrichment and design.

THE NUCLEAR FUEL CYCLE

The nuclear fuel cycle refers to those processes necessary to use uranium fuel in a nuclear power plant. The processes are illustrated on figure 5.23.

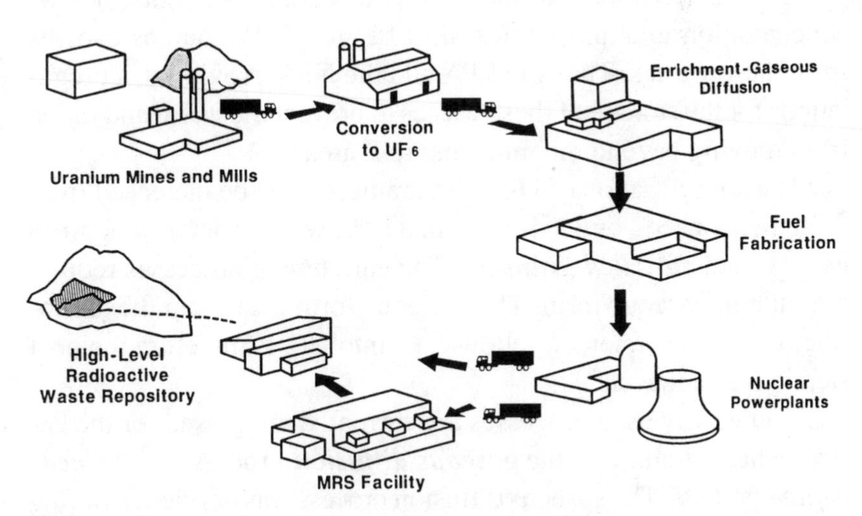

Figure 5.23 The nuclear fuel cycle. Courtesy of DOE.

The process begins with the mining of uranium. Large economical reserves of uranium exist in the western part of the United States, and smaller reserves occur elsewhere in the United States. For an ore to be economical, it must have sufficiently high concentrations of uranium to make the mining and processing competitive. Ore currently being mined in Canada is more economical than most ore in the United States, so Canada is where much of the uranium now being used in the United States comes from. Large reserves exist in Australia, several regions in Africa, and other parts of the world. There is sufficient economical uranium ore to supply the world's uranium needs for many decades to come, but not for centuries. Light water reactors require uranium-235. This translates into a requirement for large amounts of uranium ore since U-235 is only 0.7% of natural uranium. For this reason the world's low-cost uranium will eventually be consumed. This will drive the price of uranium up to the extent that an advanced type of reactor called the breeder reactor, which is described in the next section, will eventually become competitive with the present light water reactor.

A mill is generally located near the mine where the uranium ore is converted into a form of uranium oxide known as "yellow cake" because of its bright yellow color. This oxide is U_3O_8.

At both the mine and the mill, considerable amounts of low-concentration uranium wastes must be stored. Radon gas evolves from these wastes. Thus the EPA and the NRC provide strict regulations for the storage of these wastes to prevent the waste and radon from moving beyond the mine and mill areas.

The concentration of U-235 in uranium must be increased from 0.7% to about 3% or 4% for use in a light water reactor; this product is called ***enriched uranium***. The enrichment processes require uranium in gaseous form. This gaseous form is uranium hexafluoride, UF_6. Conversion of yellow cake into UF_6 is therefore the next step in the fuel cycle.

Two widely used processes are currently being used for the enrichment of uranium—the ***gaseous diffusion*** process and the ***centrifuge*** process. The gaseous diffusion process was developed during the Second World War; the first large plant to use it was at Oak

Ridge, Tennessee. This plant provided enriched uranium for the defense industry and for commercial reactors for many years, but it is now shut down. Additional plants built later at Paducah, Kentucky, and Portsmouth, Ohio, are still operating. In the gaseous diffusion method, UF_6 is allowed to diffuse through a long cascade of metal barriers. U-235 diffuses through each barrier at a slightly higher rate than U-238 so that gradually the U-235 enrichment increases as the UF_6 flows down the cascade. One of the leading scientists in the design of the gaseous diffusion plant at Oak Ridge was Manson Benedict. In 1952 he started the Nuclear Engineering Department at MIT, and in my senior year, I took his course in nuclear engineering the first time it was offered. Like Robley Evans (see page 89), Professor Benedict was an outstanding and inspiring teacher who had a great influence on my career.

The centrifuge process is in commercial use in Europe, though the process was developed originally in the United States. This process depends on a series of high-speed, revolving centrifuges. Centrifugal force causes a slight separation of U-235 from the U-238 in each centrifuge stage. The centrifuge process is more economical for a new plant, although the plants at Paducah and Portsmouth are still producing economical enriched fuel since they were built so long ago. By chance I had a personal experience with people involved in the development of the centrifuge process as well. Karl Cohen was one of the leading engineers in its early development, and later at General Electric, Dr. Cohen was the leader of the fast reactor group in which I worked during the 1960s. The inventor of the centrifuge process for isotope separation was Professor Jesse Beams of the University of Virginia, where I am a member of the faculty.

There is a third quite promising method of enriching fuel still under development called atom vapor laser isotopic separation (AVLIS). Historically, the U-235 produced for the weapons program during the Second World War was produced at Oak Ridge by yet another method using huge mass spectrographs called calutrons; the method was easier to develop rapidly than the gaseous diffusion method, but it was not as economical.

Uranium-238 is stockpiled at the enrichment plants. If the product at one end of the cascade is uranium enriched to 3% U-235, then another stream, called the ***tails***, must leave the other end of the cascade, which is depleted in U-235, just to complete the mass balance. Since the natural uranium input to the plant contains 0.7% U-235, the ***depleted uranium*** must contain less than this amount. In practice, this depleted uranium contains about 0.2 to 0.3% U-235.

The next step in the fuel cycle is conversion of the 3% enriched UF_6 to uranium dioxide, UO_2. This is followed by fabricating the oxide fuel pellets and loading them into zircaloy tubes to form the fuel rods. A number of companies that also sell reactors—General Electric, Westinghouse, ABB-Combustion Engineering, and Framatome—operate fuel fabrication plants in the United States.

The fuel remains in a reactor for a three to five-year period before being discharged, as described in the previous section. At most nuclear plants in the United States, all of the spent fuel discharged since the beginning of plant operation still remains stored underwater in the plant's spent fuel storage pool. The water cools the fuel from the radioactive fission products, which continue to decay and generate energy in the process, and the water also serves as shielding to protect workers in the area. A few plants have developed dry storage areas (for example, see figure 8.8). At these plants, the spent fuel assemblies are transferred from the pool to large steel casks after much of the fission product activity has decayed away, and the casks are stored in the dry storage area. Further cooling of the fuel is still necessary, but this cooling is accomplished by natural circulation of air past the outside of the casks.

The next step in the fuel cycle involves a choice. This choice is described in more detail in chapter 8. The spent fuel can either be stored in a permanent repository or it can be ***reprocessed***. Reprocessing means separating the fuel from the fission products so that only the fission products, together with a small amount of high mass elements like plutonium, called actinides, need to be stored. The U-235 and plutonium remaining in the fuel could be used again, or recycled. In the United States the current decision is to store the spent fuel rather than to reprocess. France, Japan, and the United Kingdom plan to reprocess. The waste to be stored is either spent fuel or fission products plus some actinides, depending on the choice.

There is an additional step being planned in the United States that will come between the fuel storage at the plant and final storage. This is called ***monitored retrievable storage*** (MRS), or interim storage. An MRS will be built somewhere to accept fuel before it is sent finally to a permanent repository, partly because this temporary storage facility will be needed before a final repository is ready.

The final stage in the fuel cycle is the permanent storage of the waste in a ***repository***. How best to accomplish this is the subject of chapter 8. The likely location of the first repository in the United States will be Yucca Mountain in Nevada.

THE CANDU REACTOR

This is an appropriate time to introduce another excellent commercial nuclear power system, designed and marketed by the Canadians and called the CANDU reactor. This is a different type of pressurized water reactor. I waited this long to talk about the CANDU reactor because it has taken this long to introduce the concepts necessary to appreciate the differences between the CANDU reactor and the ordinary light water pressurized water reactor. CANDU stands for **CAN**ada **D**euterium **U**ranium.

The moderator and coolant for the CANDU reactor is heavy water (D_2O) instead of the light water used in the PWR and BWR. Heavy water absorbs (or captures) fewer neutrons than light water. As a result of this decreased absorption, natural uranium (with its 0.7% U-235 instead of the 3 to 4% enriched U-235 used in light water reactors) can be used as the fuel in a heavy water moderated reactor. While expensive D_2O must be produced, the expensive enrichment step in the fuel cycle is eliminated.

The CANDU system is shown in figure 5.24. The reactor is composed of many horizontal zircaloy tubes filled with fuel bundles through which heavy water coolant flows. The arrangement of these tubes, or fuel channels, is shown in figure 5.25, The coolant is at high pressure, about 1500 psi (10 MPa). Heavy water is also located

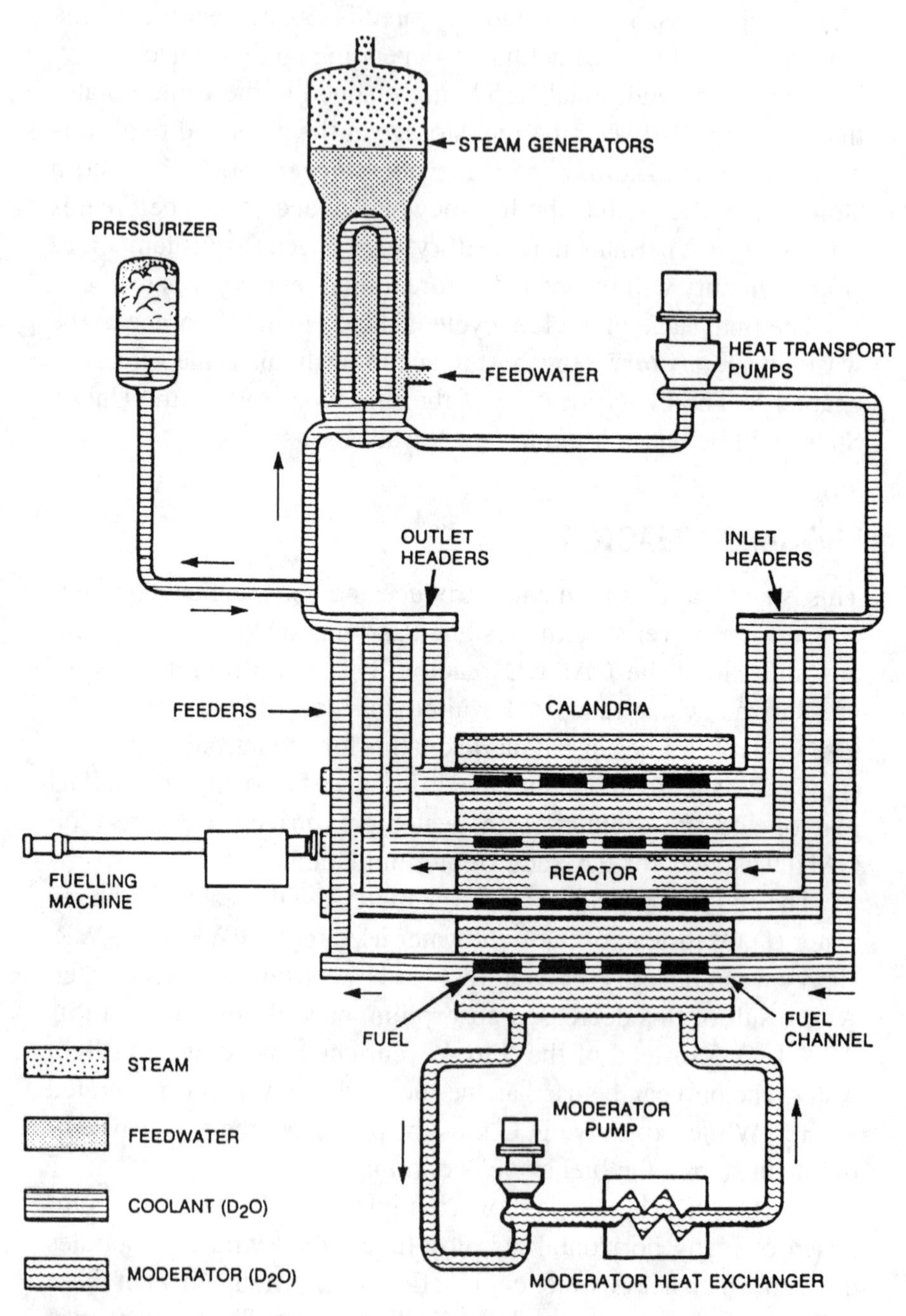

Figure 5.24 CANDU steam supply system. COURTESY OF AECL CANDU.

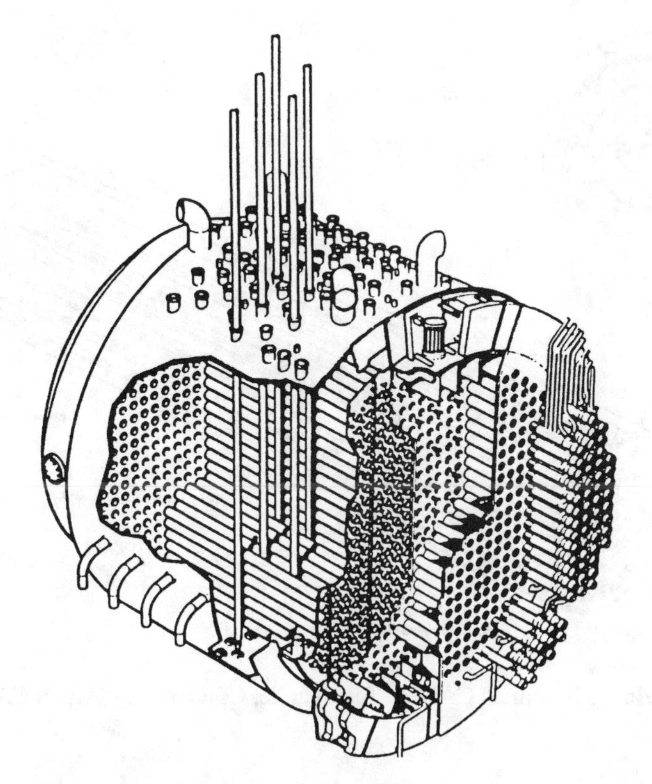

Figure 5.25 CANDU core and pressure tube arrangement. COURTESY OF AECL CANDU.

in the space between the tubes, but this heavy water is at low pressure and acts only as a moderator. The geometry is referred to as a calandria. The hot coolant from the fuel channels is sent to steam generators, where its energy is used to boil light water. This light water then flows to the turbine, like the steam generator in an LWR.

The fuel is sintered natural uranium oxide pellets clad in zircaloy and arranged in bundles, as shown in figure 5.26. The bundle shown contains 37 fuel elements and is about 4 inches (100 mm) in diameter and 20 inches (500 mm) long. A number of these bundles are placed in each fuel channel, depending on the power rating of the core. An interesting feature of this design is that the fuel can be replaced ***on line***, which means that a fueling machine pushes a new

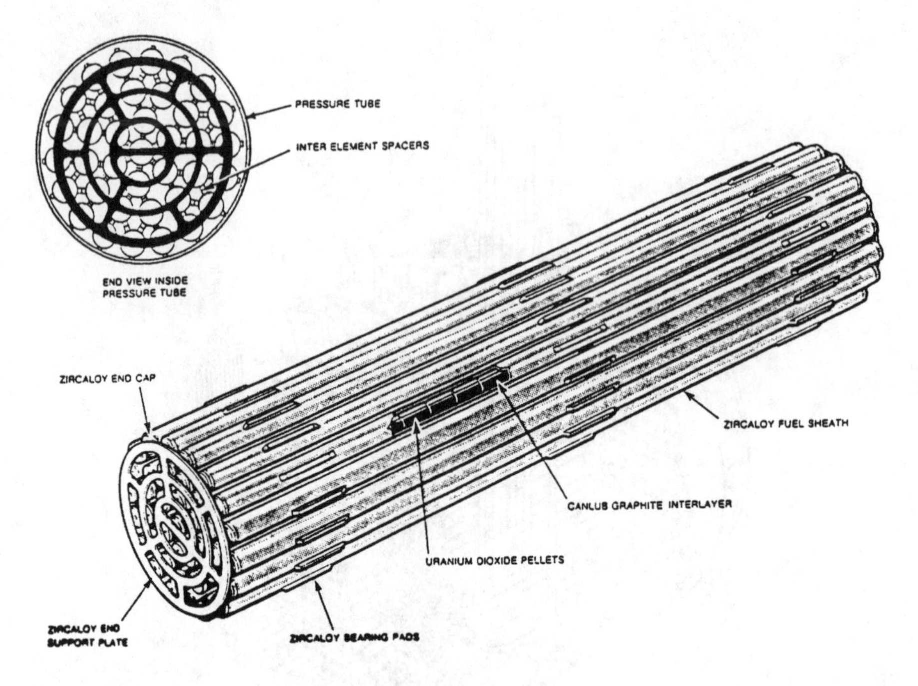

Figure 5.26 37-element CANDU fuel bundle. COURTESY OF AECL CANDU.

fuel bundle into a channel while pushing a spent fuel bundle out the other end while the reactor is still operating.

The core of a CANDU reactor is considerably larger than a light water reactor because neutrons must travel a longer distance in heavy water than in light water to slow down to thermal energies. Consequently the reactor vessel must be larger for a CANDU reactor than for a light water reactor. It would be impractical to make a large enough pressure vessel for an entire CANDU core. The solution is to keep the coolant in high-pressure tubes, which can be made easily with thin tube walls. The reactor vessel, or calandria shell, has to contain only the low-pressure moderator. The CANDU reactor also has containment and emergency core cooling systems similar to LWRs.

THE BREEDER REACTOR AND PLUTONIUM

Earlier in this chapter, under the discussion of nuclear reactions, it was noted that uranium-238 could capture a neutron and produce plutonium-239. It was also noted that Pu-239 is like U-235 in that it is an excellent fuel. In some reactor designs it is possible to use plutonium as the primary fuel and simultaneously produce plutonium from U-238. The question arises, is it possible to design a reactor that produces more plutonium than it consumes? That would be remarkable because if it were so, the U-238, generally considered of little value, would suddenly take on great value.

Indeed, production of more plutonium than is consumed is possible. A reactor designed to accomplish this feat is called a ***breeder reactor*** because a net quantity of fuel (plutonium) is produced. Most breeder reactor designs are fast reactors, the reactor type discussed earlier for which the term "fast" refers to the neutron speed, or energy. The anthropomorphical analogy is carried further when the U-238 is called a ***fertile*** isotope since it gives birth to the Pu-239.

It is possible to make a thermal reactor breed, and some experimental thermal breeders have been built and operated. They use thorium as the fertile isotope. Uranium-233 is bred from thorium-232, and U-233 is another excellent fuel material, like U-235 and Pu-239. Thermal breeders have not proven to be as promising as fast breeders, however, and these designs have been abandoned.

The breeder reactor represents an inexhaustible source of energy. There is enough uranium-238 to last forever in a breeder reactor economy—or at least for hundreds of thousands of years and, as far as I am concerned, that is forever. Together with solar, fusion, and geothermal energy, fission energy is one of the four inexhaustible sources of energy. It is the closest one to being economically competitive for the generation of electricity. Sometimes, in media stories, the breeder reactor is mistakenly omitted from this list, and concern is expressed that fission energy is limited since there is a limited amount of economical uranium to fuel the present light water type of reactors. On the scale of many decades, or a century, the light water reactor is indeed limited, as discussed in the preceding section. However, a breeder reactor can burn roughly 60 times as much of the natural uranium mined as a light water reactor by converting the U-238 to Pu-239. For this reason, the cost of the uranium is inconsequential for a breeder reactor. There is now enough uranium stored at the former uranium enrichment plant at Oak Ridge, Tennessee, which is depleted of U-235 and therefore worthless for thermal reactors, to last the United States for hundreds of years if all of our electricity were made from breeder reactors—without mining any more uranium at all. For the longer term, it would be economical to get uranium from sea water to fuel a breeder reactor; there is just about an infinite amount of uranium in the oceans. The amount in the oceans is always increasing since soil is continually being carried by streams and rivers into the oceans, and most soil contains uranium.

6

Current Nuclear Plants and Why They Are Safe

All reactors contain large quantities of radioactive fission products that must be isolated from the environment. Reactors also contain radioactive elements of mass higher than uranium, such as neptunium, plutonium, and americium. These elements are called ***transuranics***, and they too must be isolated. The presently operating reactors, those designed between 1955 and 1980, have numerous ***engineered safeguards***, safety features engineered into the design, which protect the public from the release of radioactive materials. These safeguards, together with a general discussion of reactor safety, are described in this chapter.

The present nuclear power plants throughout the United States, Western Europe, and the eastern countries of Japan, Korea, China, and Taiwan are nearly all of similar light water reactor designs, with common safety features and philosophy. They are very safe and pose virtually no threat to the health and safety of the public. There is a "margin of safety" built into every reactor design that keeps the probability of an accident that can affect the safety of the public sufficiently low. Even so, much of U.S. society believes that the margin of safety in present reactors is insufficient. It is indeed possible to design reactors with greater safety margins than those existing in plants designed before 1980. Construction of such nuclear plants is currently underway in Japan and will likely eventually take place in the United States. Designs of some of these plants are described in chapter 7. It is necessary, however, to understand the designs of the pre-1990 generation of plants, together with their engineered safeguards, in order to understand and appreciate the remarkable advances in design that are offered by the new 1990 designs.

ENGINEERED SAFEGUARDS

There are four basic barriers between the radioactive material in a reactor and the surrounding environment, as illustrated in figure 6.1. The first is the fuel itself. The UO_2 fuel is a ceramic that contains the solid fission products and the transuranics. Only the gaseous fission products such as xenon, krypton, and, at fuel temperatures, iodine can escape the fuel pellets as long as the fuel remains solid. The

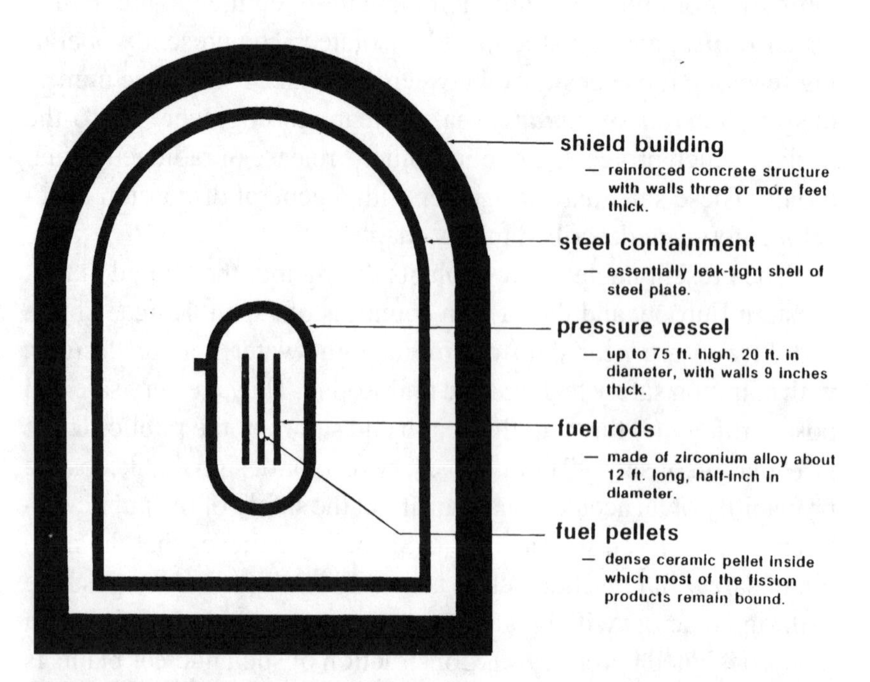

Figure 6.1 Defense in depth; four barriers between fission products and the environment. COURTESY OF NUCLEAR ENERGY INSTITUTE.

second is the fuel rod. The zircaloy cladding contains the fission product gases and prevents contact between water and the fuel. Third, there is the primary coolant system, so that leakage of radioactive materials other than the inert gases xenon and krypton into the water coolant is contained in the primary system. The fourth barrier is the containment, which is a large leak-proof building or structure (or "leak-tight," as we say in the nuclear industry) that would contain any fission products, including gases, which might escape from the fuel and the primary system in the event of an accident.

Containment

In the next few paragraphs I will use the general term ***containment*** to include a large thick-walled concrete structure and a leak-tight steel shell, or vessel, inside this structure. For recent BWRs, the word "containment" more accurately applies only to the steel shell; the concrete structure is called the ***shield building*** (as shown in figure 6.1). For PWRs the term ***containment building*** is used for both the concrete building and the steel shell inside the building; the steel shell is called the containment vessel, or sometimes simply the containment. On the other hand, in the everyday jargon of the nuclear industry, all these terms are frequently shortened to simply "containment," and I will do this too.

The containment serves several purposes. First, if a primary system water pipe ruptures, or a pipe entrance or exit nozzle on the reactor vessel fails, the containment must contain the steam that is produced as the high-pressure water flashes to steam. Second, the containment keeps any radioactive materials that might escape from the primary system inside the containment and, thus, keeps them away from the environment and the public. Third, the containment must protect plant personnel from the radiation that might escape from the primary system during an accident. The first and second functions are performed by a leak-tight steel shell, or liner, on the inside of the containment structure, together with an additional structure called a pressure suppression pool in the case of the BWR. The third function is performed by the thick concrete walls of the containment structure, or shield building.

Pressurized Water Reactor Containment

There are two main types of containment in use for the PWR. The first and most widely used is the large, dry containment, illustrated in figure 6.2. This is the large, cylindrical building that one often sees in pictures of nuclear power plants (as in figure 5.13). This structure, usually with a domed top, is of the order of 300 feet tall and 170 feet in diameter, with only about half of its height above ground level. The walls of the building are made of concrete, three to four feet thick, reinforced throughout with steel bars. These walls provide both missile protection (say from a ruptured turbine blade or a telephone pole being hurled by a tornado) and shielding. If the plant is near an airport, the walls may be required to withstand an airplane crash. The bottom of the containment is a concrete pad 10 to 15 feet thick for earthquake protection. A leak tight steel liner is located inside the concrete walls. The steam generators and the entire primary coolant system are inside the containment; the turbine and rest of the steam system are outside the containment in a separate building.

The steel liner inside the walls must hold the steam pressure from a pipe break and remain leak tight to contain any fission products that may have escaped from the fuel. The containment atmosphere is initially at atmospheric pressure or, in a few designs, at subatmospheric pressure—about two-thirds of an atmosphere. A tremendous amount of energy resides in the high-temperature, high-pressure primary water of an LWR. If a primary system pipe breaks, the water will flash to steam as it "blows down" into the low-pressure containment building. The containment must have sufficient volume to contain this steam without exceeding the design pressure limit of the steel liner. Typical design pressures are 60 to 80 psi, which are greater than the maximum steam pressure that can be generated during blowdown. Large tests of containment building designs conducted during the 1980s at the Sandia National Laboratories demonstrated that the containments can hold pressures at least as high as, and generally considerably higher than, the design pressures. All actual plant containments are periodically tested to ensure their ability to contain high pressure with an acceptably low leak rate.

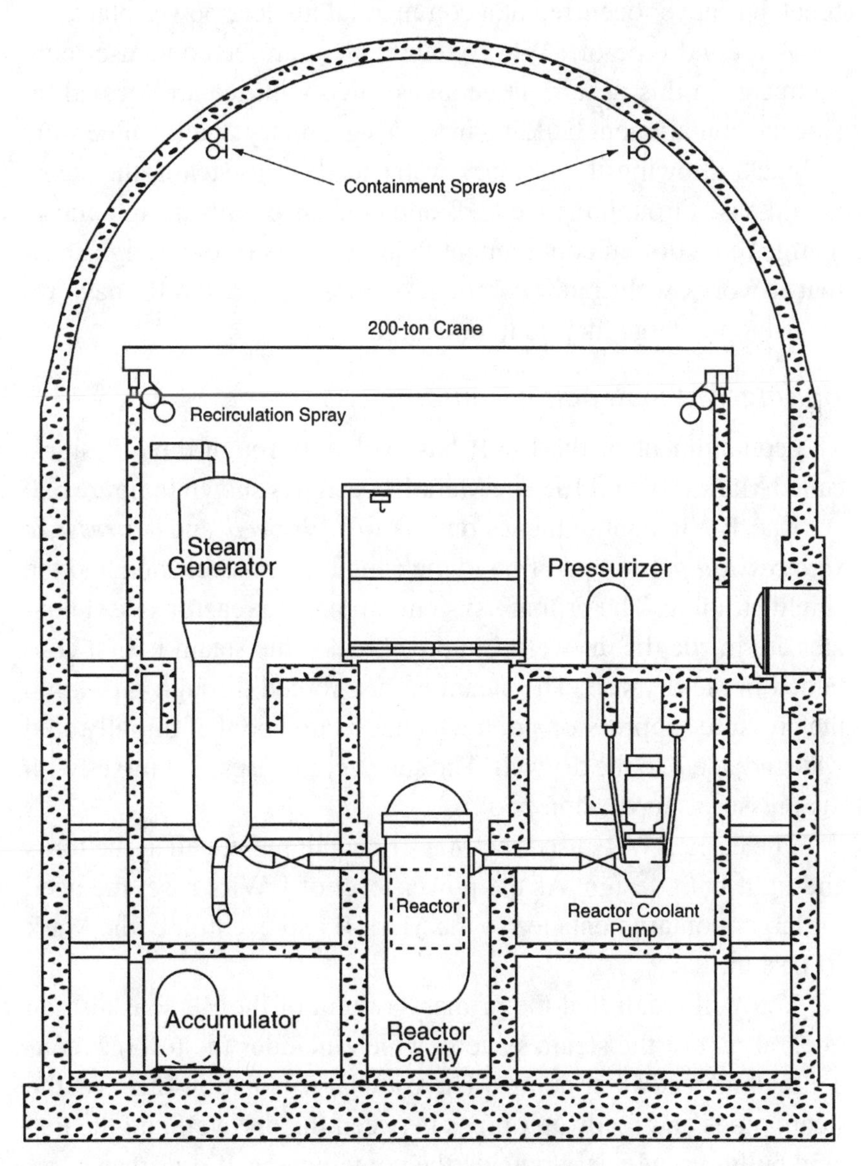

Figure 6.2 PWR containment; large, dry containment design.

These containments have never been tested in an actual large loss of coolant pipe break accident because such a primary system pipe break has never occurred at a commercial nuclear power plant.

A second type of PWR containment is an ice condenser containment. In this design, large areas filled with ice are located inside the containment building in such a geometry that if a pipe were to break, allowing the primary water to flash to steam, the steam would pass through the ice beds and condense. This design allows a large reduction in containment volume. Tests of the design show that it works well, but, nevertheless, most recent PWRs have returned to the large, dry containment design.

Boiling Water Reactor Containment

The containment of the BWR has evolved through three designs, called Mark I, II, and III. The Mark III design is shown in figure 6.3.

The BWR containment consists of a ***drywell*** and a ***pressure suppression pool*** in addition to the containment steel shell and the shield building. The primary system piping and reactor vessel nozzles are inside the drywell. If a pipe breaks, the steam is first contained in the drywell. The steam is then forced through pipes into the pressure suppression pool, which is a large pool of initially cold water adjacent to the drywell. The steam condenses as it mixes with the pressure suppression pool.

The early BWRs used the Mark I containment, called the torus and light bulb design. As the power level of BWRs rose, the need for larger containments led to the Mark II and eventually the Mark III design.

You will recall that the primary system of the BWR is also an integral part of the steam system, which includes the turbine, condenser, and feedwater pumps. There are large isolation valves in the steam pipes and feedwater pipes between the dry well and the turbine building, which is outside the containment. If a primary system pipe ruptures, the isolation valves immediately close in order to ensure that any fission products from the reactor core remain inside the dry well and pressure suppression pool.

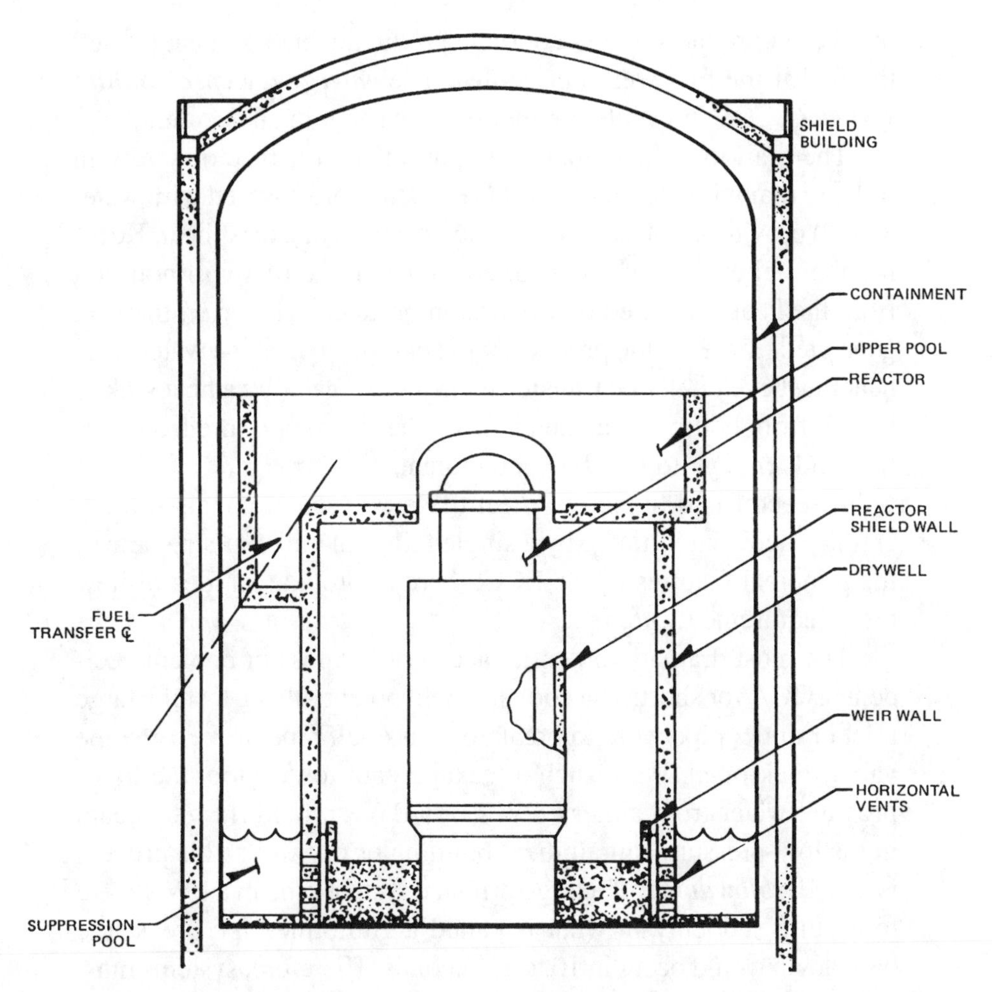

Figure 6.3 BWR Mark III containment and shield building. COURTESY OF GENERAL ELECTRIC COMPANY.

Emergency Core Cooling System

In an accident such as a pipe break in an LWR, one or more of several mechanisms would operate to shut down the neutron chain reaction in the core, thus terminating the fission process. These mechanisms include a decrease in pressure leading to boiling of the water, insertion of control rods, and the addition of boron to the water. Although heat generation from fission would quickly cease, the fission products would continue producing enough heat to melt the fuel if the fuel were not cooled. The ***emergency core cooling system*** (ECCS) prevents the fuel from melting in an accident.

There are two fundamental requirements for reactor safety in order to maintain fuel integrity: (1) keep the core covered with water and (2) provide a heat sink, i.e., a way to get rid of excess heat. Keeping the fuel covered with water ensures that the fuel will not melt from the heat generated by the fission products. However, the heat generated by the fission products will continue to heat the water. This heat must eventually be transferred from the water by a heat sink to keep it from boiling away and to keep the steam pressure from getting high enough to fail the containment.

A reactor is designed to meet these two requirements for any credible accident initiated by a single failure. I will explain some of the principal features of the ECCS here to provide an idea of how these requirements are met.

The most dramatic potential accident is a loss of coolant accident, LOCA for short. The coolant could potentially be lost if a large inlet or outlet pipe were to break, or if a small pipe broke, or some valve stuck open. Were such a leakage path to develop, the high-pressure water from either a PWR or a BWR would flash to steam in the low-pressure containment building or dry well. This process is called a ***blowdown***. If a large primary system pipe in a PWR were to rupture suddenly in what is called a guillotine pipe break, the blowdown would occur in 10 to 15 seconds. Therefore, systems must respond quickly and automatically if such an event occurs.

Since this accident is considered credible, every light water reactor is equipped with an ECCS to keep the core covered and provide a heat sink in such an event. This system ensures that additional supplies of water are available to keep the core covered in the event of a pipe break and that an ultimate heat sink is available when the heat sink system used for normal shutdown is lost.

The ECCS in a PWR consists of several components, as illustrated schematically in figure 6.4. First there is an ***accumulator*** on each primary system loop. (Recall that there are generally three or four loops in a plant.) This is a large tank of borated water under about 700 psi pressure from nitrogen present in the upper part of the

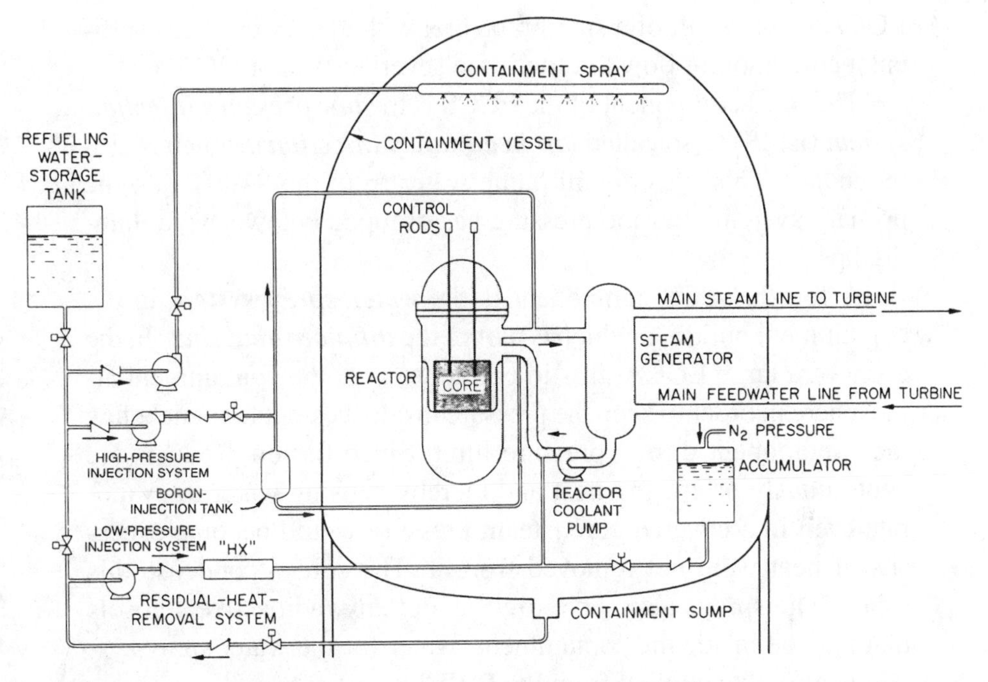

Figure 6.4 PWR emergency core cooling system (ECCS).

tank. If the pressure in the reactor vessel drops below 700 psi, which would happen in a large pipe break, the accumulators would automatically discharge their borated water into the reactor vessel. During normal operation, the accumulators are isolated from the primary system by check valves, which are valves that allow water to flow in one direction (in this case from the accumulator to the primary system) but prevent water from flowing in the opposite direction.

A second component is the ***high pressure injection system*** (HPIS) (also called ***high head safety injection***). This system includes a pump and an outside source of water (the refueling water storage tank, RWST), which can be pumped into the primary system at high pressure. There are ***two redundant*** HPIS systems. Redundancy means that either of the two systems can supply sufficient emergency core cooling to keep the core covered so that failure of one of the systems does not prevent core cooling. As long as the pressure remains high, which might be for a considerable time if the

LOCA is the result of a small pipe break, the HPIS provides sufficient core cooling flow to reduce the severity of the LOCA.

The third component of the ECCS is the ***low pressure injection system*** (LPIS) (also called ***low head safety injection***), which is also redundant. This system will pump water from the RWST into the primary system after the pressure has dropped below several hundred psi.

A fourth ECCS component is the ***water spray system*** in the containment building, which is part of the ***ultimate heat sink.*** In the event of a large LOCA, it is important to cool the containment atmosphere in order to keep the pressure inside the containment below the containment design pressure limit. Since fission products are continuously producing heat and thereby causing water to evaporate even in a covered core, steam pressure would become excessive if heat were not removed from it. The sprays condense this steam. The sprays also serve to wash out any radioactive aerosols that may be inside the containment. Water for the water spray system is initially supplied from the RWST.

A fifth component, and another part of the ultimate heat sink, includes heat exchangers that remove heat during long-term core cooling. Long-term core cooling begins when the RWST is empty and the ECCS is aligned to pump the water that has collected in the containment sump back into the primary system. This water is at an elevated temperature since it has absorbed heat from the reactor core. It is therefore pumped through heat exchangers (labeled "HX" on figure 6.4) that are cooled by the plant cooling water system. The heat is ultimately transferred to the environment from the plant cooling water system.

A sixth component, not shown in figure 6.4, is the backup electrical supply system. This system consists of ***emergency diesel engine electric generators*** that are needed to drive the ECCS pumps in the event of a failure of normal electric power. Normal power comes from a different power station and is called ***off-site power***. Power generated by the diesels is referred to as ***on-site power***. Although only one diesel generator is needed, all nuclear plants have at least two diesel generators to provide a redundant power source in case one of the diesels does not start up. Batteries are used to start up the diesels.

The external source of water for the ECCS is the RWST, discussed in chapter 5 in the section on fuel management. This is a large tank of water needed for flooding the area around and above the reactor vessel during normal refueling of the reactor. By the time this water is exhausted during an emergency, water would be spilling over into the sump (the low spot) inside the containment. This water in the sump would then become the water for recirculation through to the ECCS components.

The ECCS for the BWR is somewhat similar to that of the PWR though there are differences due to differences in the reactor and containment designs. Similar redundancies are required for the BWR as for the PWR.

NUMBERS AND PROBABILITIES: IF REACTORS ARE SAFE, THEN HOW SAFE?

Concepts of probability add greatly to an understanding of nuclear reactor safety. Unfortunately, many people have difficulty understanding probabilities. (Parenthetically, let me add that I've heard that even mathematicians have trouble with numbers sometimes; a mathematician recently told me that there are three kinds of mathematicians, those who can count and those who can't!) But seriously, without a feeling for numbers it is difficult to obtain a perspective about reactor safety, about how safe nuclear energy really is compared to other means of generating electricity or to other industries with which we are rather familiar. Helping students develop a sense of the implications of relative probabilities, or the ability to compare risks with benefits, is an important challenge for secondary school science teachers in our technological society.

Probabilistic Risk Assessment

Nuclear engineers calculate the probabilities of all possible accidents that might happen at a nuclear plant and the resulting probability that members of the nearby public might be harmed by such an accident. These are scientific calculations, based on well-established methods—methods that are taught in graduate school courses and are verifiable by experience. The discipline has a name; it's called

probabilistic risk assessment, or PRA for short. It was developed first for aircraft and space accident analysis, but it has been used extensively for nuclear safety analysis since the middle 1970s. The results are not exact numbers, as provided by mathematics, or the accurate figures involved in engineering design, but the results are dependable estimates based on scientific evidence. During the past few years, teams of engineers have calculated these probabilities for every nuclear power plant in the United States. The calculated probability that an accident severe enough to cause core damage may be, for example, one chance in 100 000 per year. This means that the actual chance is probably somewhere between 30 000 and 300 000 per year. The actual probability is almost surely somewhere between one in 10 000 and one in 1 000 000 per year. This is the calculated value divided by or multiplied by 10, which we say is within a factor of 10 of the calculated result. This is useful in comparing the safety of different methods of generating electricity with each other or with some standard for what is acceptable. We will see later (chapter 9) that the probabilities of harm to the general public from producing the same electricity from coal are considerably greater than from nuclear. ***In fact, if our society applied the same safety standards to the generation of electricity from coal that we require for nuclear power, the cost of generating electricity from coal would be so high that it would effectively prohibit the use of coal for electricity***, which is our main source of electricity and the main use of coal in the United States today.

Before I present actual accident probability results, it is instructive to gain some understanding of the methodology used in probabilistic risk analysis. My colleague, Professor Reed Johnson, has provided a clear explanation of the concepts by examining what might happen in an automobile accident. Suppose you are asked to estimate the probability that you will crash into a car that stops suddenly in front of you. The probabilistic risk analysis is illustrated in figure 6.5, which starts with the initiating event and is followed by an ***event tree***. The first probability is the probability that the event will be initiated, i.e., that a car will suddenly stop in front of you; this probability is P_{ie}. The first branch in the event tree is whether

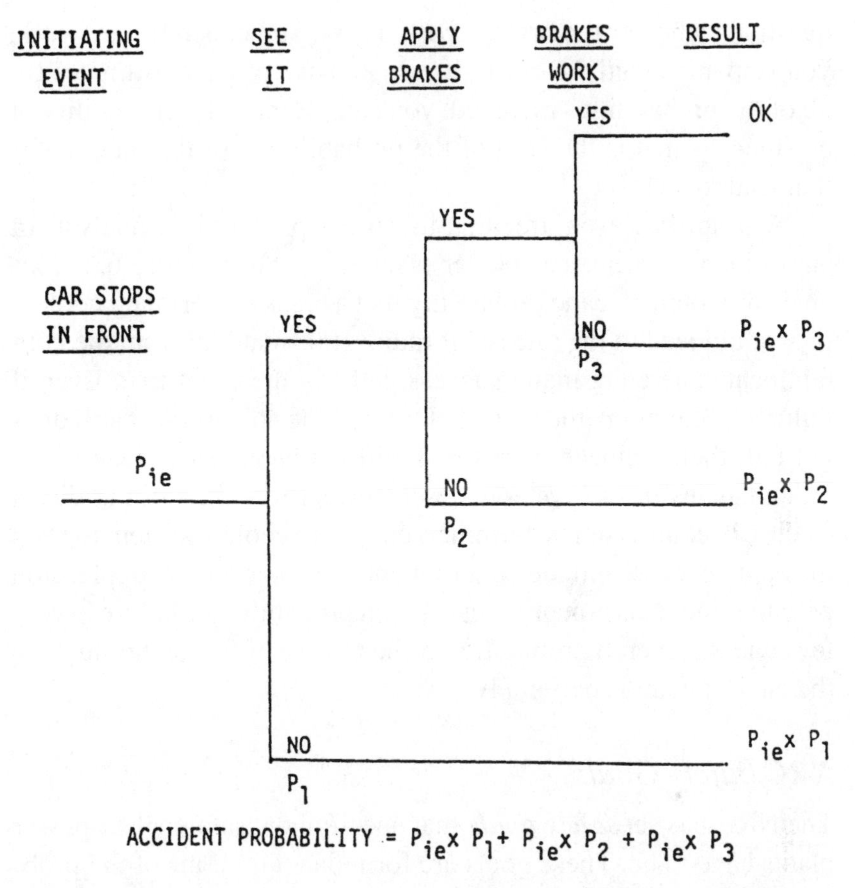

Figure 6.5 Event tree example for probabilistic accident analysis of an automobile accident. To obtain the final results listed, this example assumes that the probabilities P_1, P_2, and P_3 are all much less than 1. Thus, the "yes" branch probabilities [for example, $(1\text{-}P_1)$] are essentially equal to 1.

you see that the car has stopped. There is a probability associated with this branch; the probability is P_1 that you do not see that the car has stopped. In that case you will crash. The probability of the accident occurring this way is the product of the two probabilities, P_{ie} times P_1. If you see the car stop, you follow the "yes" branch. The next question is whether you apply the brakes in time. If you don't, you crash. If you do, you proceed to the next branch, which questions whether the brakes work. If not, you crash. If they work, you stop and avoid the crash. If research can provide estimates for all of the probabilities involved, you can estimate the probability of crashing, which is the sum of the probabilities for the three paths that lead to a crash.

A simplified event tree that illustrates probabilistic analysis of an accident in a nuclear power plant is shown in figure 6.6. This analysis would give the probability that a break in a primary coolant pipe could lead to the release of radioactive material from the containment. There are many success paths in the event tree. Even if failures occur along the way, if at the end the containment still does not fail, then radioactive material will not be released. Due to the redundancies in the systems described earlier—i.e., in the diesel electric backup systems, the emergency core cooling system to keep the core covered, and the containment spray system or suppression pool to remove fission products—each probability of failure is very low, and the overall probability of the release of fission products to the environment is extremely low.

NRC Safety Goals

The NRC has set ***safety goals*** that specify how safe nuclear power plants have to be. These goals are formulated in terms of the probability of a human fatality resulting from a nuclear accident. There are two goals:

1. The probability that a person living near a nuclear power plant will die soon after a nuclear accident from radiation released in the accident must be less than 0.1% of the total probability that a person will be killed in any accident.

2. The probability of death from cancer for any member of the public following an accident must be less than 0.1% of the total probability that a person will die of cancer from all causes.

Numerically, the first safety goal means the following. The average probability per year that a person will die from all accidents is about one chance in 2000. Therefore, the first safety goal means

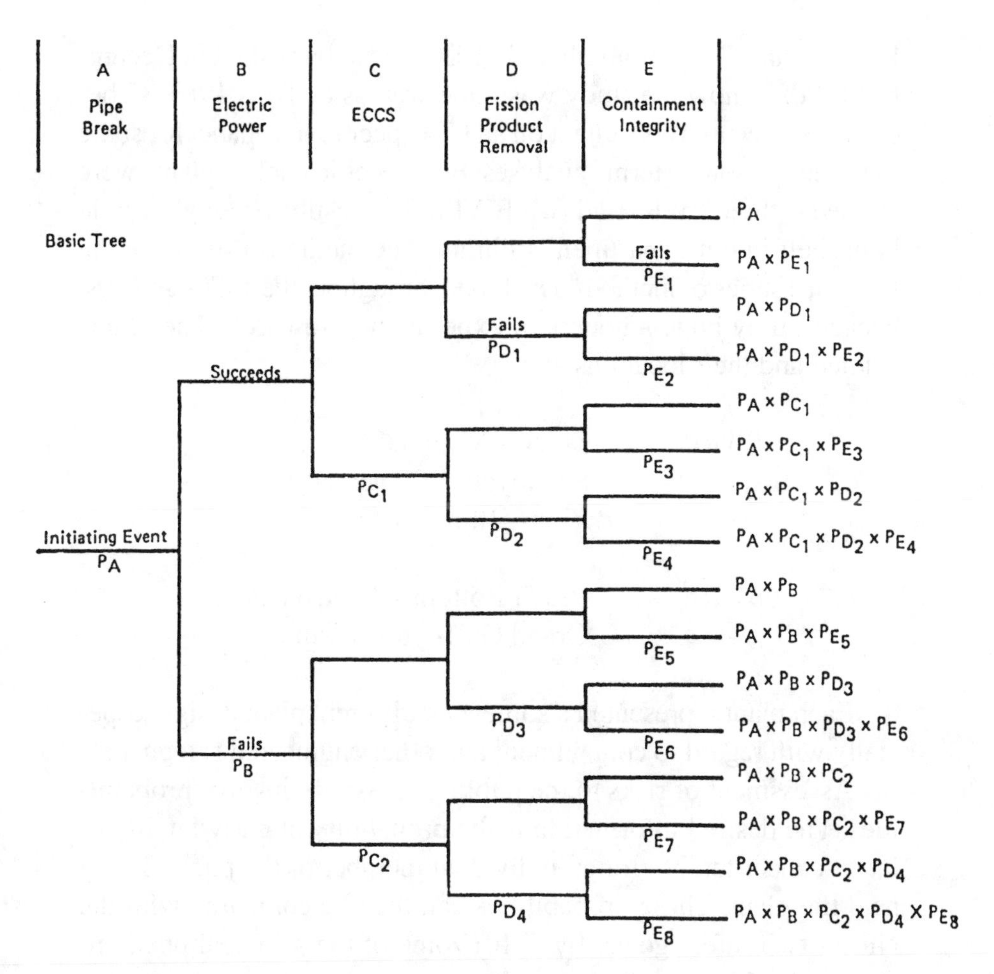

Figure 6.6 Simplified event tree for a coolant pipe break.

that the probability per year that the person living next to a nuclear plant will die soon after a nuclear accident from radiation released in the accident must be 1000 times less than the above probability, or less than one chance in two million.

Severe Accident Risk Study, NUREG-1150

The most extensive study of reactor safety ever conducted was completed in 1990 by the NRC. The study is summarized in the report "Severe Accident Risks: An Assessment for Five U.S. Nuclear

Power Plants," report number NUREG-1150, published in December 1990. I know the study was underway as early as 1984-85 because I served as a consultant on an NRC peer review panel for some of the early "source term" analyses. Five specific nuclear plants were studied—three PWRs and two BWRs. The results occupy a whole bookshelf in our department's library. The methods used were the latest in a series of methods developed throughout the 1970s and 80s, backed up by huge amounts of experimental research. The plants studied and their locations are:

PWRs:	Surry—Virginia
	Sequoyah—Tennessee
	Zion—Illinois
BWRs:	Peach Bottom—Pennsylvania
	Grand Gulf—Mississippi

Each plant represented a somewhat different plant design, especially with regard to containment and other engineered safeguards.

Assessment of risks to the public necessarily involve probabilities. The desired information is the probability of early fatality or latent cancer fatality for an individual member of the public living near the plant. These probabilities can then be compared with the NRC's two safety goals. I will list some of the principal phenomena involved in the calculations in order to provide a perspective on the complexity of such an assessment.

The analysis is broken down into four fundamental parts:

1. The frequency of core damage.
2. Radioactive source term inside the containment.
3. The probability of containment failure.
4. Calculated off-site consequences.

Core damage refers to accidents that will damage the fuel, and here we mainly mean fuel melting, with the consequent release of fission products, especially iodine, cesium, and strontium, into the water coolant. There are numerous scenarios, or sequences of events, that have the possibility of causing core damage. Each of these must be analyzed in order to add up the total probability that core damage might occur.

For the analysis of core damage, the potential causes of accidents are divided into two categories:

- Internal events.
- External events.

Internal events are accidents initiated by failure of equipment internal to the plant. External events are accidents initiated by earthquakes, fire, volcanoes, lightning, and aircraft crashes.

The main internal events are:

- Station blackout—Loss of all off-site and on-site alternating current sources of electricity (i.e., all electricity other than stored direct current power from batteries).
- ATWS (pronounced at-wus)—Anticipated transient without scram, meaning a transient in the reactor system in which the control rods fail to scram the reactor so the power level is not reduced to the fission-product decay-power level.
- Loss of coolant accidents (LOCA).
  - Large pipe breaks in which system pressure is suddenlly reduced.
  - Small LOCA, examples of which are a small pipe break, a stuck open valve, and a seal failure on a reactor coolant pump.
- Containment bypass—meaning the opening of a flow path from the containment to the environment by a mechanism such as an inadvertently opened valve to a line from the containment, a check valve failure in the PWR ECCS, an isolation valve failure in the steam line of the BWR, or a containment penetration failure.
- Steam generator-tube rupture for a PWR.
- Main steam-line break for a PWR.

The only two external events that have an important impact on risk are earthquakes and fires. These external initiating events may lead to internal failures listed above and to containment failures.

The probability that these events can lead to core damage is calculated from a probabilistic risk assessment of the reactor system. The results are not single numbers but rather a range, or distribution, of probabilities. Results are reported in NUREG-1150 as probabilities per year of causing core damage.

The first result that I will quote from NUREG-1150 is the probability of core damage. I will quote this result for only one of the reactors analyzed—a PWR at the Surry Nuclear Power Station in Virginia. Results for the other reactors are similar. The average, or mean, probability of core damage per year from all potential internal accident scenarios is 4×10^{-5} per year. Inverting this number says that there is one chance in 25 000 per year that a core damage accident will occur at Surry. Next we consider the range of calculated probabilities. The analysis says that there is a 95% certainty that the probability of core damage is not greater than 1.3×10^{-4} per year, or that the probability is not greater than one chance in 8000 per year. The other end of the probability range says that there is a 5% chance that the probability is as low as 7×10^{-6} per year, or that the probability may be no greater than one chance in 140 000 per year.

Oh me! you might be thinking by now. Why can't scientists and engineers simply come up with a single number? Or better yet, just tell us whether something can or can't happen? The answer is that the best scientific safety estimates are always ranges of probabilities (which, of course, is one reason why it's so hard for scientists to communicate with the public).

I remind you at this point that the probabilities of core damage quoted above do not translate into the probabilities of damage to the public. Not yet. We still have to determine the probability that radioactive material will be released from the containment before we can talk about potential health effects on the public. The probability is high that even core damage will have no adverse health effect on the public. Three Mile Island was a case in point, where core damage was extensive, but there was almost no release of radioactive materials.

Next we must consider the second factor listed above, the radioactive source term. This refers to the amount of radioactive material that can get out of the fuel, into the water coolant system, and then into the containment. Here we are particularly interested in the fission products iodine-131, cesium-137, and strontium-90. Also of interest are radioisotopes of the fission products tellurium, lanthanum, ruthenium, and barium. Models for transport pathways from the fuel, through the primary system, and into the containment are available for this calculation for each fission product.

The next step in the analysis is containment failure analysis. Several things can happen following core damage. First, the containment may not fail at all, as was the case at Three Mile Island. Second, the containment may fail in such a way that some fission products may escape into the environment. It makes a difference to the risk to the public whether the failure is early in the accident or late, e.g., several days, after the start of the accident. Another possible path is bypass of the containment, which actually occurred in the early part of the Three Mile Island accident. There is even a further possibility for some of the BWR designs and that is purposely to vent the containment for some accident scenarios.

The final step is the calculation of the off-site consequences of these potential releases of radioactivity. These results depend on several things such as the probabilities for various weather conditions that govern how far and fast the radionuclides travel, the surrounding population density, the extent and timing of evacuation of this population, and the damage to health due to exposure to the various radionuclides that reach the people. Here the linear, no-threshold hypothesis (see chapter 2) is used to link cancer risk to radiation exposure.

Significant Results of NUREG-1150

I have selected two results that I think best illustrate the significance of these exceedingly elaborate analyses, if you can imagine summarizing years of study into a few simple results. The first is a comparison of the probability of an individual fatality with the NRC safety goals that were given on page 160. The second is the likelihood of an accident at a nuclear power plant large enough to cause at least one early fatality among the general public.

The results of the comparison with the NRC safety goals are given in table 6.1. The first line is the average probability per year of the death of an individual soon after an accident (early fatality) near one of the PWRs (Surry) and one of the BWRs (Peach Bottom). The second entry in the table is the average probability per year that a person will eventually die of cancer as a result of an accident at one of the plants. These results are compared to the NRC safety goals, which are listed in the third column in the table.

The results in table 6.1 clearly show that, for both the PWR and the BWR, ***the calculated probabilities for damage to the public are way below the safety goals set by the NRC***. These results hold for all the PWRs and BWRs analyzed and for the range of probabilities calculated in addition to the average results. The results are also valid for externally initiated accidents.

The next significant result is the probability of a large enough release of radioactivity to cause at least one early fatality to the public. The average probabilities vary between reactors from 10^{-6} to 10^{-9} per year for the different LWRs. These numbers translate to the following summary result from NUREG-1150—one which perhaps can be remembered from all of this massive study—that ***the likelihood per year of an accident at a nuclear power plant large enough to cause at least one early fatality to the public is in the range of one in one million to one in one billion per year***. That's mighty safe.

TABLE 6.1 Comparison of Early and Latent Fatality Risks with the NRC Safety Goals For Internally Initiated Accidents at Specific PWRs and BWRs (from "Severe Accident Risks," NUREG-1150)

	Surry (PWR)	*Peach Bottom (BWR)*	*NRC Safety Goal*
Average probability of an individual early fatality per year	2×10^{-8}	5×10^{-11}	5×10^{-7}
Average probability of an individual latent cancer per year	2×10^{-9}	4×10^{-10}	2×10^{-6}

NUREG-1150 is the best and most comprehensive safety analysis that's been done in the nuclear reactor field, and it is grounded on a huge amount of experimental research data. Keeping in mind the results quoted above, you can see why it is difficult for an engineer to give a precise answer to the question, "Are reactors safe?" He or she is apt to respond, "How safe do you think is 'safe'?" I generally answer yes, reactors are safe, right off the bat, because I think they are plenty safe. But what I really know are the results like those I have just quoted, which are the products of decades of safety-analysis development and elaborate calculations. The results italicized in the two preceding paragraphs indicate enormous safety.

School science teachers understand numbers and probability. My experience in teaching week-long courses in nuclear energy for science teachers is that after learning about the low probability of nuclear accidents due to the many safeguards built into plant designs, they understand why nuclear energy is so safe, especially compared to other methods of generating electricity. I have had public school science teachers tell me that they are aghast at having for years shown students antinuclear films—with all the potential horror stories but no perspective on the likelihood of these events ever happening—because these were the only films that happened to be in the school's library. I hope the readers of this book will try hard to use the probabilities discussed here to gain a perspective on how safe reactors are. Further perspective on risk from nuclear energy is provided in chapter 9.

THREE MILE ISLAND

Well then, you naturally ask, what about Three Mile Island? Did the ECCS work at Three Mile Island? The answer is yes. Then what went wrong?

What happened was that, first, a feedwater pump failed. The auxiliary feedwater pump did not work because the line had been valved off by an operator by mistake, in violation of procedures. Operator error. The loss of feedwater caused a steam generator to boil dry, with a resultant partial loss of heat sink for the primary coolant system. This caused the primary system pressure to rise to the extent that a pressure relief valve at the top of the pressurizer opened, exactly as it was supposed to do. When the pressure was reduced, the valve did not close—a second equipment failure. Continued loss of coolant through the relief valve triggered the ECCS. The high pressure injection system (HPIS) pumped water into the primary system to make up for the water being lost out the stuck valve, again exactly as it was supposed to do. Then the major error occurred. The operators misunderstood what was happening and manually turned off the HPIS—another operator error. This led to uncovering and melting of a large fraction of the core. Had the HPIS not been turned off, core damage would have been completely avoided. After the damage was done, the engineers and operators figured out that the relief valve had stuck open and that water was being lost, so they turned the HPIS back on. This stopped the accident. Early in the accident a small amount of fission product gases—xenon, krypton, and iodine—escaped with some water that bypassed the containment until a valve on a pipe leading through the containment was closed. After that no more fission products escaped from the containment. The amount that escaped early was not enough to cause any harm to the public.

Invoking a familiar cliche, we might say that Three Mile Island was a wake-up call for a young maturing nuclear industry. After Three Mile Island a number of changes were made in all plants to make them safer. More important, the level of operator training was vastly improved. A national Institute of Nuclear Power Operations (INPO) was established by the nuclear utilities to upgrade and oversee operator training and reactor operation nationwide. Since that time a worldwide organization called the World Association of Nuclear Operators (WANO) has been created because it is recognized that a nuclear accident can affect people across national boundaries and have an impact on nuclear energy everywhere.

WHY A CHERNOBYL-TYPE ACCIDENT CANNOT OCCUR IN A WESTERN-TYPE REACTOR

The reactor that exploded at Chernobyl was a design thoroughly different from the light water reactors used commercially outside the old Soviet Union. It was moderated by graphite, with boiling water cooling the fuel in pressure tubes between the graphite. There was no containment. There was a pressure suppression pool that could help in some kind of accidents, but it was irrelevant for the accident that occurred. The fundamental difference was that the reactor was unstable at low power. The "power coefficient" at low power for a Chernobyl-type reactor is positive. The power coefficient for western light water reactors is negative at all power levels. (The power coefficient at Chernobyl was negative at normal operating power, but not at low power.) A negative power coefficient is stable; when the power is raised by withdrawing a control rod, the power rises and then levels out automatically. When the power coefficient is positive, however, raising the power introduces changes that make the power continue to rise even faster. If the power is raised too rapidly, the chain reaction can get out of control, causing the power to rise too rapidly to be stopped by the control rods.

This is what happened at Chernobyl. The operators were running an experiment at low power. They violated many operating procedures in order to run the experiment (for example, disconnecting many control rods, which would be inconceivable in western plants), and then they raised the power rapidly from low power. With its positive power coefficient, the reactor became supercritical, the power rose too fast to control, to about 1000 times normal operating power, for a long enough time to produce enough high-pressure steam to blow the reactor apart and blow the 100 ton steel cover clear off the top of the reactor. This is the explosion people refer to. This simply cannot happen in commercial reactors outside the former Soviet Union. No reactor can be licensed with a positive power coefficient in western countries. Even with operator errors, one cannot overcome the stabilizing effect of the negative power coefficient inherent in all western reactor designs.

Unfortunately, there are many Chernobyl-type reactors operating throughout the former Soviet Union. The Soviet Union sold reactors to eastern European countries, but these were all pressurized water reactors with negative power coefficients at all power levels. No Chernobyl-type reactors were ever built outside of the Soviet Union. The Soviet-designed PWR reactors in Eastern Europe are not built to western standards, however, and there is widespread concern throughout the western nuclear industry that some of the PWRs in Eastern Europe are not safe enough.

RADIATION EXPOSURE TO THE PUBLIC FROM NORMAL REACTOR OPERATION

Let's step back from accidents now and consider radiation exposure to the public during the ***normal*** operation of a nuclear power plant. A very small amount of radioactivity is released to the environment during the normal operation of a nuclear plant. Generally, the amount is less than the amount of radioactivity released in the smoke from an electric power plant that uses coal as a fuel since there is much thorium and uranium in coal. That's not the main message of this paragraph, however, as interesting (and to some, surprising) as this may be. The message of this paragraph concerns the amount of radioactivity from nuclear plants. The NRC sets a limit on the amount of radioactivity released during the normal operation of any nuclear power station by setting a limit on the amount of radiation that any person outside the plant can receive from this release. This limit is spelled out in federal law. The law says that a person living at the boundary of the plant can receive a dose of no more than 5 mrem in any one year from the radioactivity released from the plant. The NRC assumes that a person at the boundary of the plant will eat a certain number of fish from the water next to the plant, drink milk from cows that graze next to the plant, breathe air for some fraction of the day next to the plant, etc. The average dose absorbed per person in the United States from natural background and medical radiation in one year is 350 mrem, as discussed in chapter 2. The hypothetical additional 5 mrem for the person living next to the plant is clearly small compared to the national average.

A few scientists have tried to link cancer and infant mortality to radiation received from the normal operation of commercial nuclear plants and other nuclear facilities. Such sensational reports always make the front page of the newspapers. Each time this happens, scientists from the national laboratories have to respond with extensive analyses to show why the statistics used by the alarmists are invalid, and invariably this is clearly shown—but, of course, that is not "news." There is simply no way that the low levels of radiation from normal operation can have any measurable effect on the public.

However, since sensational reports do exist, the U.S. Congress mandated a two-year study a few years ago by the National Cancer Institute (NCI) to determine if there was any evidence of excess cancers in the populations around nuclear facilities in the United States. (I recall when this study was announced. It was front page news. The mere fact that such a study was ordered, I am sure, made many think that nuclear plants must be unsafe; you know—where there's smoke there's fire.) In 1990, NCI published its findings in the report, "Cancer in Populations Living near Nuclear Facilities." ***There was no evidence whatever of any excess cancers among people living near nuclear facilities***. The study covered 52 nuclear power plants in operation before 1982 (the earliest starting in 1957) and 10 other nuclear facilities, including enrichment, fabrication, and fuel reprocessing plants. Nine of these other facilities were defense plants.

I have concentrated on safety at nuclear power stations in this chapter. You might well ask, if operation at nuclear power plants is so safe, then what about uranium mines, mills, and other parts of the nuclear fuel cycle? Maybe the NRC and the nuclear industry have forgotten about them, and they're unsafe, with all their radon and whatever! The fact is that the NRC and the nuclear industry thoroughly evaluate every part of the commercial nuclear enterprise, not just the power plants. All facilities can be and will be made as safe as the NRC and society demands. However, to discuss every facet of these issues would make this long chapter even longer. I do discuss the part of the nuclear fuel cycle that appears to be of most concern to the public—high-level nuclear waste disposal—in chapter 8. So let's move on at this time to the exciting new reactor designs now on the horizon and learn about their automatic safety features.

7

The New Designs: Simplicity and Automatic Safety

Despite the safe designs of the present U.S. commercial nuclear reactors, there are many people in our society who demand more safety and many who are still just plain afraid of nuclear energy. For this reason, the engineers and planners responsible for reactor design made the clear decision to introduce major changes in reactor designs that would make them even safer. During the twenty years since the last nuclear plant order by a U.S. utility, nuclear engineers have successfully developed these new designs. The new light water reactor designs are not only safer but also economically competitive with the earlier designs and with other methods of generating electricity.

Two classes of new reactors have been developed—***evolutionary plants*** and ***passive plants***. Evolutionary means that the designs have evolved as a safer modification of the present large nuclear plants. Passive means that safety systems operate automatically, powered by gravity or natural convection rather than mechanically active equipment. While both new reactor types are remarkable, most of this chapter will be spent describing the passive designs because their safety features differ from the present plants more dramatically than the evolutionary plants. The first section, however, will be devoted to the evolutionary designs since some of them are being built at this time.

Some comparisons of power costs between the evolutionary and passive designs and other methods of generating electricity are given at the end of this chapter.

EVOLUTIONARY DESIGNS

Evolutionary designs will be competing with passive designs when new nuclear plants are ordered in the United States. The main two evolutionary plants being designed in the United States are the advanced PWR called System 80+ of ABB Combustion Engineering and the advanced BWR (ABWR) of General Electric. Westinghouse is working with Mitsubishi of Japan on another evolutionary PWR. In addition, the French and Germans have formed an international company called Nuclear Power International (NPI), and they are designing a 1450 MWe evolutionary nuclear plant called European Pressurized Water Reactor, EPR.

These reactors are large in size—similar to the largest of currently operating reactors. They have a significantly greater margin of safety than present reactors, but the changes are not as dramatic as those in the passive designs. The General Electric design was developed jointly with Japanese companies. One General Electric ABWR began operation in Japan in 1996; a second will start up in 1997. The power level of the ABWR is 1300 MWe. The power level of ABB Combustion Engineering's System 80+ is 1350 MWe. Three plants similar to System 80+, called System 80, have been built at a nuclear power station in Arizona, two are operating in Korea, four more are under construction there, and the System 80+ has been chosen as the basis for the next generation of plants in Korea.

Both the GE ABWR and the ABB Combustion Engineering System 80+ have been approved by the NRC and are expecting final certification in 1996. With NRC certification, only site approval is required to obtain an operating license, which means that the utility or generating company can be assured that the reactor will be licensed by the NRC if the site selected is qualified.

The evolutionary designs include substantial modifications of the current large nuclear plants for greater simplicity and added safety. These evolutionary designs incorporate the following:

- Simplified and more rugged systems.
- Safety systems that are more diverse and redundant.
- Additional features to mitigate consequences of a severe accident.
- State-of-the-art computer and data processing technologies.
- A more operator-friendly control room.
- Special design attention to ease of maintenance.
- Designs that are more economical.

The probabilities of core damage and harm to the public are reduced below those of present commercial light water reactors by more than a factor of ten. Simplified design has resulted in factory instead of field assembly of many components, leading to a reduction of construction time to four years, as demonstrated by the first two ABWRs built in Japan.

Many important changes were developed for ABB Combustion Engineering's System 80+. One is the adoption of a steel spherical containment that is enclosed in a cylindrical concrete shield building, as illustrated in figure 7.1. This design provides the safety advantage of dual containment and incorporates specific features to reduce consequences of severe core-damage accidents.

The refueling water storage tank in the system 80+ has been moved to inside the containment for greater reliability of emergency core cooling. An alternate electric power source (a combustion turbine) has been added to the emergency diesel generators to provide backup electricity. All emergency core cooling systems have been improved for greater safety. For example, one of the emergency systems consists of four independent trains, two powered by electric motors and two by steam turbines that don't require electricity for operation. The addition of these and numerous other safety features has reduced the probability of core damage for the System 80+ to a factor of 10 below the U.S. nuclear utility goal and various international goals and a factor of 100 below the NRC's goal.

An important change in the advanced BWR design is that the reactor coolant recirculation pumps are now installed ***inside*** the reactor pressure vessel, as illustrated in figure 7.2. In earlier designs,

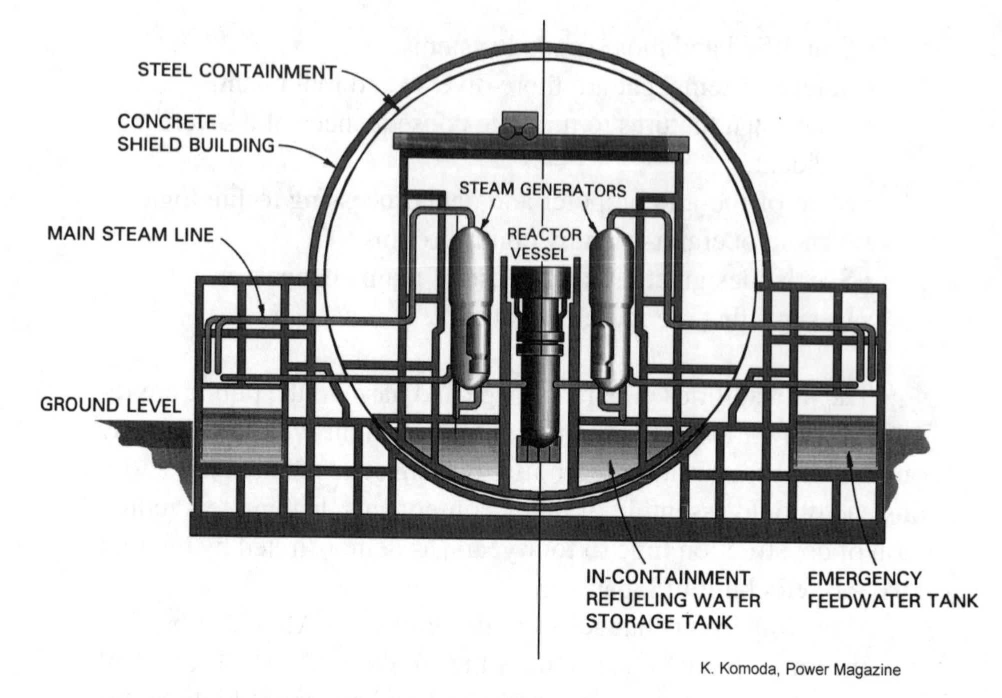

Figure 7.1 Steel spherical containment and cylindrical shield building for ABB Combustion Engineering's System 80+. COURTESY OF ABB COMBUSTION ENGINEERING AND K. KOMODA, *POWER ENGINEERING MAGAZINE.*

the recirculation pumps were external to the vessel (see figure 5.15) so that the piping and inlet and outlet nozzles were susceptible to breaks that could initiate a loss of cooling accident. The motors for the internal pumps are outside the vessel, but the actual pumps are inside so that the possibility of rupture of pipes and large nozzles is eliminated.

Simplification and added redundancy have greatly reduced the chance for core damage in the ABWR. The ABWR has three completely independent and redundant divisions of safety systems that are mechanically and electrically separated, with an emergency diesel generator for each division. One of the high-pressure backup cooling systems is powered by reactor steam instead of electricity in order to provide protection in case of loss of all electricity. The

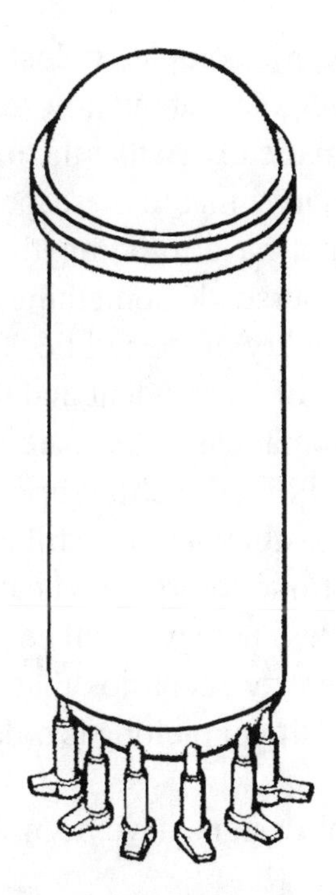

Figure 7.2 Reactor coolant recirculation pumps in General Electric's evolutionary Advanced Boiling Water Reactor. COURTESTY OF GENERAL ELECTRIC COMPANY.

systems have the ability to keep the core covered with water at all times in an accident.

In the ABWR, plant response has been fully automated so that operator action is not required in a loss of coolant accident for three days, which is the same as for the passive SBWR design.

PASSIVE DESIGNS

Now we turn to the automatically safe passive designs. First we will discuss the passive light water cooled reactor designs. These will be followed by gas-cooled and liquid sodium cooled reactor designs.

These designs are indeed remarkable. Automatically safe means that in the event that everything goes wrong—for example, a pipe breaks, all electric power is completely lost, and the operators suddenly forget all they ever knew about how to operate the plant—the plant shuts down automatically so that the fuel and the plant are not damaged and no radioactive material escapes into the environment. The operators would not have to do anything, at least for several days. They would, of course, do something, but they don't have to in order to ensure the safety of the public. The plant automatically shuts down in the event of an accident and the decay heat from the fission products is automatically dissipated successfully from the fuel without damaging it. Fuel melting becomes virtually impossible. I say *virtually* impossible; the fuel still contains radioactive fission products that continue to produce heat, so melting can never be absolutely impossible. There will still be probabilistic risk analyses that attempt to quantify scenarios that could possibly lead to melting. But the probability of melting is orders of magnitude below today's reactors.

Five passive reactor designs have been or are being developed, as listed below.

1. AP600 (Advanced Passive—600 MWe)—a pressurized water reactor.
2. SBWR (Simplified Boiling Water Reactor—600 MWe).
3. PIUS (Process Inherent Ultimate Safe—640 MWe)—a pressurized water reactor.
4. GT-MGR (Gas Turbine-Modular Gas-Cooled Reactor).
5. ALMR (Advanced Liquid Metal Cooled Reactor).

How these designs work is the subject of the rest of this chapter. We will make use of what was learned in chapter 6 in order to appreciate the differences in the newer designs that make them so safe.

These automatically safe designs are based on the natural laws of physics—things like gravity and natural convection—which never fail. In some of the designs, absolutely no ***active*** components or equipment have to function for the reactor and the public to be protected; the needed safety systems are entirely ***passive*** and, hence, operate automatically. Active systems are those that require some mechanical or electrical action in order to function. Passive systems have no moving parts and thus require no mechanical changes in order to function. Some of the new designs still have valves that must open in emergencies, and, hence, have active components, but the valves are either air-operated valves that automatically open with the loss of electricity, or they are battery-operated valves that rely on stored energy rather than on an active system to provide the electricity. In none of the designs are pumps needed for safety, and it is pumps that require electricity. In none of the designs do valves needed for safety require actively generated electricity for their operation.

In none of the passive designs is electricity needed to protect the reactor or the public (other than stored electricity for the operation of valves in a few cases). In none of the designs is operator action needed to shut the reactor down or to remove the decay heat to keep the fuel from melting.

Two features of the passive designs are their smaller size and, for some, their modular design. These features allow much faster construction times, an important factor in reducing plant costs. The new water reactors have a power rating of about 600 MWe. The accompanying reduction in complexity of design and size of the components allows more fabrication and assembly off-site and this reduces the time required to construct the plant.

The new gas-cooled and liquid metal cooled plants are ***modular*** in design. This means that the plant will consist of several small reactors, each called modules, which can mostly be fabricated off-site at a central manufacturing plant. A utility or power generating company can order a single module rather than a complete plant and can add modules later as the need for electric generating capacity grows.

In the past, the general rule of economies of scale applied to electric power plants, meaning that as the plant size increased, the cost per kilowatt-hour was reduced. This led to large plants, as high as 1450 MWe in France. The future use of smaller plants must bring into bear other features to offset this principle of economy of scale. These features, including less complexity, more off-site assembly, and shorter construction time, and the fact that often utilities do not need such large capacity additions to their grids at one time, have made the smaller designs attractive. The evolutionary designs remain somewhat more economical than the smaller size passive designs, however, so it still remains to be seen which reactors will be more popular in the United States when new plants are ordered.

In order to appreciate the remarkable advances in reactor safety that have been introduced into the designs of the 1990s, it is necessary to understand the concept of fluid flow by natural convection. Therefore, before describing the new designs, we diverge temporarily for a brief tutorial on natural convection.

NATURAL CONVECTION

Natural convection is the phenomena that causes the chimney in your fireplace to work, so many of you are already somewhat familiar with the concept. Consider the flow loop illustrated in figure 7.3. Suppose the fluid in the loop is heated at a low point and cooled at a high point. The fluid in the vertical pipe on the left (the hot leg) will be at a higher temperature than the fluid in the pipe on the right. The density of a hotter fluid is lower than the density of a colder fluid, which means that the hotter fluid is lighter than the colder fluid. As a result of this density difference, the hotter fluid will tend to rise, while the colder fluid will tend to fall. This causes the fluid to flow around the loop—by the laws of nature rather than by a pump. This flow is what we mean by natural convection. This is in contrast to forced convection in which a pump would be used to force the flow.

In your chimney, the air, gases, and smoke from the fire are hot and want to rise up the chimney. The air heated in the fire is replaced by cold air in the living room. The higher the chimney, the greater will be the pressure differential that drives the flow and the greater will be the air flow rate.

Mathematically, the flow is forced by the pressure difference between the hot leg and the cold leg. The hydrostatic pressure at the bottom of the vertical leg in figure 7.3 is the product ρgh, where ρ is the fluid density, g is the gravitational constant, and h is the height of the fluid. Hydrostatic pressure is what you feel on your ears

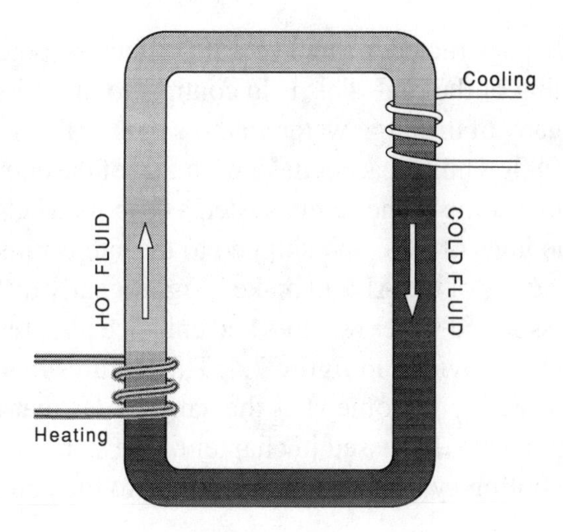

Figure 7.3 Natural convection loop.

when you swim to the bottom of the deep end of the swimming pool—as you all know, the greater the depth, the more pressure you feel. Since the density of cold fluid is greater than that of hot fluid, ρgh is greater for the cold leg than it is for the hot leg. This creates a pressure difference that causes the fluid to flow in the direction shown. The greater the height, h, the faster the flow, just as a fireplace with a tall chimney will draw better than one with a short chimney. One might wonder why the speed of the flow doesn't just get faster and faster, since there appears to be a force imbalance. The reason is that the flow produces friction between the fluid and the pipe and at some flow velocity the drag from this friction exactly compensates for the difference in hydrostatic pressure. Thus the speed of the flow is self limiting.

Now back to a description of the new designs.

AP600, AN ADVANCED PASSIVE PWR DESIGN

The 600 MWe AP600 pressurized water reactor was designed by Westinghouse. Its certification by the US Nuclear Regulatory Commission is expected in the late 1990s. Westinghouse claims that the

AP600 can be constructed in three years from the first pouring of concrete to loading of the fuel. This is in contrast to times longer than a decade for many of the large water reactors built in the United States during the 1980s. Unlike earlier designs, many of the equipment systems, including some of the safety systems described below, are fabricated as modules off-site and shipped to the site for installation.

Many features of the AP600 make it significantly different from previous pressurized water reactors and enhance its safety. A sketch of the AP600 is provided in figure 7.4. The features that add safety are described below in somewhat the same order that engineered safety features were discussed in chapter 6. We start with the emergency core cooling system and then proceed to the containment.

AP600 Emergency Core Cooling

It would be great for a reactor to be designed to accommodate all possible accident scenarios. As noted in chapter 6, early reactors were designed with engineered safeguards for protection against accidents. The early, or conventional reactors, however, can encounter problems in certain extremely low probability accidents, such as a sequence of events that combines a pipe break with the loss of all off-site and on-site electricity and all heat sinks, which might lead to failure of the containment. Having said this, I hasten to remind you once again that none of these sequences of events has ever occurred in the 40-year history of commercial PWR or BWR reactor operation—not even when operator error caused fuel melting at Three Mile Island. Can we build economically competitive reactors that make the likelihood of containment failure even from these accidents vanishingly small? The answer is yes, and that is precisely what the new designs do.

Unlike earlier designs, the AP600 has been designed to accommodate ***all*** accident scenarios. The AP600 can accommodate any accident resulting from a single failure with no operator action at all. The most important accidents that would get earlier designs into trouble are listed below, starting with the most serious. This list is followed by a description of the remarkable methods that the AP600 uses to handle these accidents.

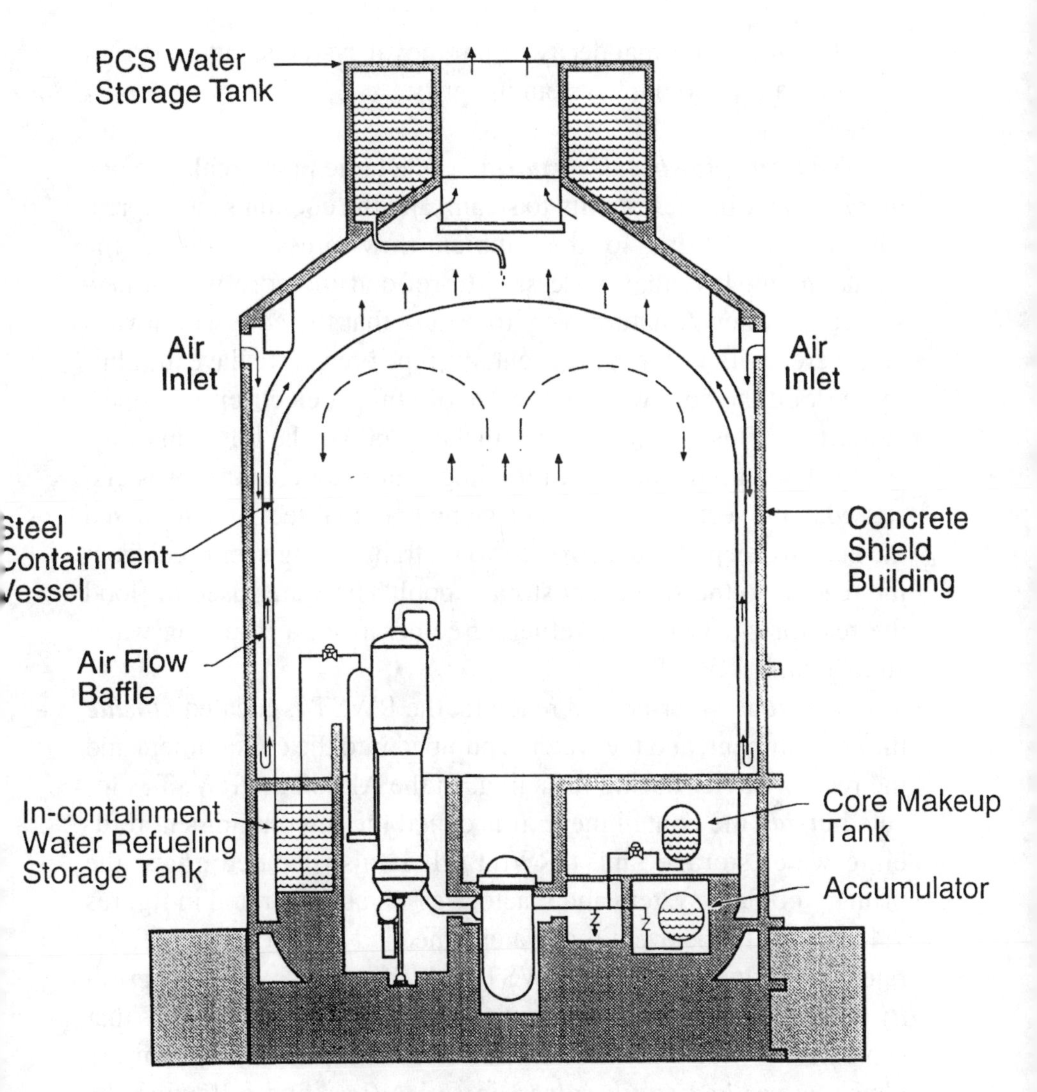

Figure 7.4 Westinghouse's AP600 design, showing containment design features and sources of water for emergency core cooling. COURTESY OF WESTINGHOUSE ELECTRIC CORPORATION.

- Large pipe break and loss of all electricity and normal decay heat removal systems.
- Small leak or small pipe break and loss of all electricity and normal decay heat removal systems.

- Loss of all normal decay heat removal systems, but no loss of coolant or coolant system integrity.

The large pipe break scenario. In a large pipe break, the primary coolant flashes rapidly to steam, and the coolant system pressure quickly drops to the ambient low pressure inside the containment. It is then necessary to reflood the core with a new source of water. A natural way to ensure that the core will always be covered with water in the event of a pipe break is to have the ability to flood the core without the use of pumps, electricity, or operator action. This is accomplished in the AP600 in the following way.

In chapter 5 the need for refueling water storage tanks was discussed in the section on fuel management. In that section I discussed the need to keep the fuel covered when transferring spent fuel from the reactor to the spent fuel storage pool. The water used to flood the reactor cavity during refueling comes from a refueling water storage tank (RWST).

In current water-cooled reactors, the RWST is located ***outside*** the containment, and the water is pumped into the containment and the reactor cavity during refueling. In the AP600 the RWST is located ***inside*** the containment. It is called the in-containment refueling water storage tank (IRWST). It is also located ***above*** the primary coolant system and reactor vessel, as illustrated in figures 7.4 and 7.5. In the event that water is needed to cover the core, borated water flows from the IRWSTs into the reactor vessel ***by gravity*** through pipes and valves as shown in figure 7.5. The valves that separate the water in the IRWSTs from the primary system are ***check valves*** and ***squib valves,*** in series (i.e., one following the other). Water can flow in only one direction through check valves—from the IRWSTs to the primary system. These valves are represented by the large Zs with the flow-direction arrows beside them in figure 7.5. The squib valves, shown just below the check valves in the figure, are one-way action valves that are activated by an explosive charge when needed to function in an emergency. Thus during normal operation, high-pressure water from the primary system cannot flow into the IRWSTs. On the other hand, if the pressure in

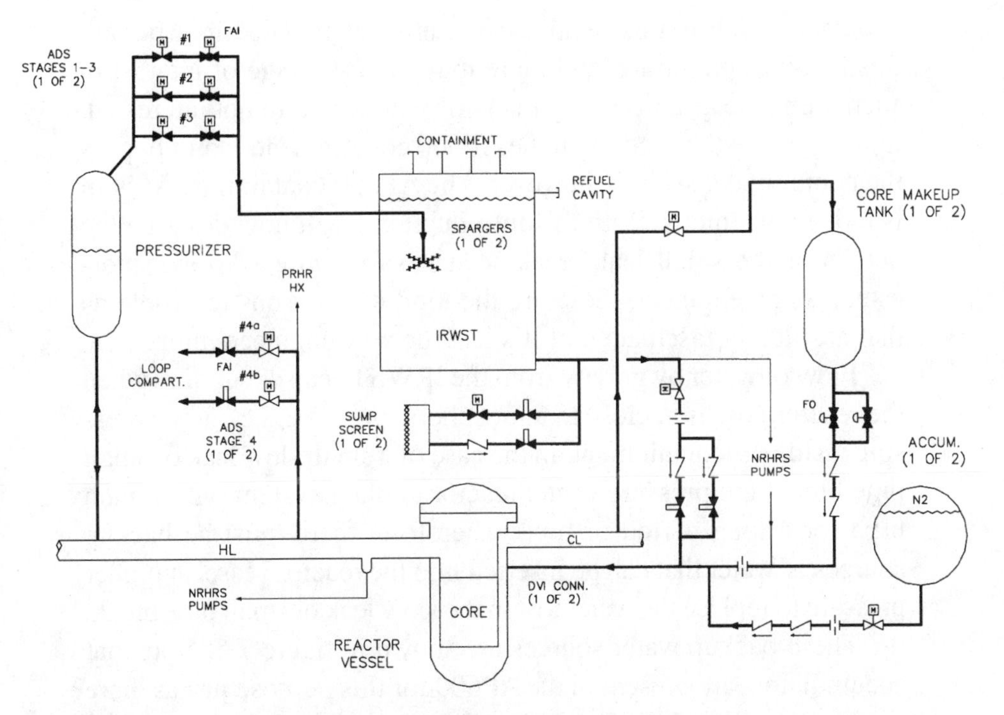

Figure 7.5 AP 600 passive emergency core cooling system. COURTESY OF WESTINGHOUSE ELECTRIC CORPORATION.

the primary system drops to the containment pressure, as in a large pipe break, the water will flow by gravity from the IRWSTs to the primary system, thus flooding the core. Therefore, pumps are not required to force the water to flow from the IRWSTs to the core; hence, keeping the core covered in the event of a pipe break is not dependent on the need for electricity. Enough water is in the tanks ***to cover*** the ***entire*** primary system completely, pipes and all, so that the core can never be uncovered.

The small leak or small pipe break scenario. We are now moving on to the next, but less serious, accident. If we can accommodate the above accident, it should be easier to accommodate this one, and it is. I wanted to go into detail about the above accident because I wanted to make obvious to you how the AP600 can handle that worst of all accidents. I don't know whether you want to know about the details of this next less serious accident scenario. After all, this involves plumbing, and there may be a shortage of interest in such mundane affairs among many of you. If you're not interested, jump ahead to the section on the AP600 containment. Don't bypass the containment section, however. The AP600 containment system is indeed ingenious. With this introduction I will now describe the details of the small leak/break scenario for those who are interested. As an engineer, these are the kind of solutions to problems that absolutely fascinate me! It's just the way engineers think.

Flow of water by gravity from the IRWSTs can occur only when the reactor core is at close to atmospheric pressure, i.e., at the pressure inside the containment. In the case of a small slow leak or small pipe break, the pressure in the reactor coolant system will remain high for a long period of time. Therefore, there must be backup sources of water that can be inserted into the reactor vessel at higher pressure to replace the water lost in the slow leak or small pipe break.

These backup water sources are shown in figure 7.5. Note that accumulators are present in the AP600 for this purpose just as there are in conventional PWR designs. Recall that these are passive devices that use nitrogen under pressure rather than pumps to force borated water into the reactor vessel if the primary system falls below the nitrogen pressure. There are also two core makeup tanks connected to the cold legs that contain borated water, which can be inserted directly into the reactor vessel at even higher pressure through a system of redundant air operated valves. If the water inventory in the primary system is depleted by a slow leak or small pipe break and if the main coolant pumps are not working due to a loss of electric power, water from these tanks will automatically flow into the reactor vessel by gravity, even at high pressure. Hence, the water flow is again independent of the need for pumps and electricity. If the main coolant pumps are still working, makeup water from these tanks can still flow into the reactor vessel.

Water from the refueling water storage tanks will be needed after the water from the accumulators and the core makeup tanks is depleted. As we saw in the accident scenario in the previous section, water will flow by gravity from the IRWST only if the primary system is at the ambient containment pressure. Thus while the accumulators and CMTs are emptying, the pressure in the primary system is reduced through a series of automatic depressurization system (ADS) valves and pipes that discharge primary water into the IRWST through the spargers. After the primary system has been depressurized, the water in the IRWST is available to flow by gravity into and over the reactor vessel and primary system, as was the case in the above large pipe break scenario.

Loss of normal decay heat removal system scenario. A passive decay heat (or residual heat) removal heat exchanger (PRHR) is located in the IRWST that can remove decay heat in the event that the reactor coolant system remains leakproof, but all normal decay heat removal systems fail to operate. This is illustrated in figure 7.6. One side of the PRHR is connected to the cold side of the primary system at the lower channel head of the steam generator; the other

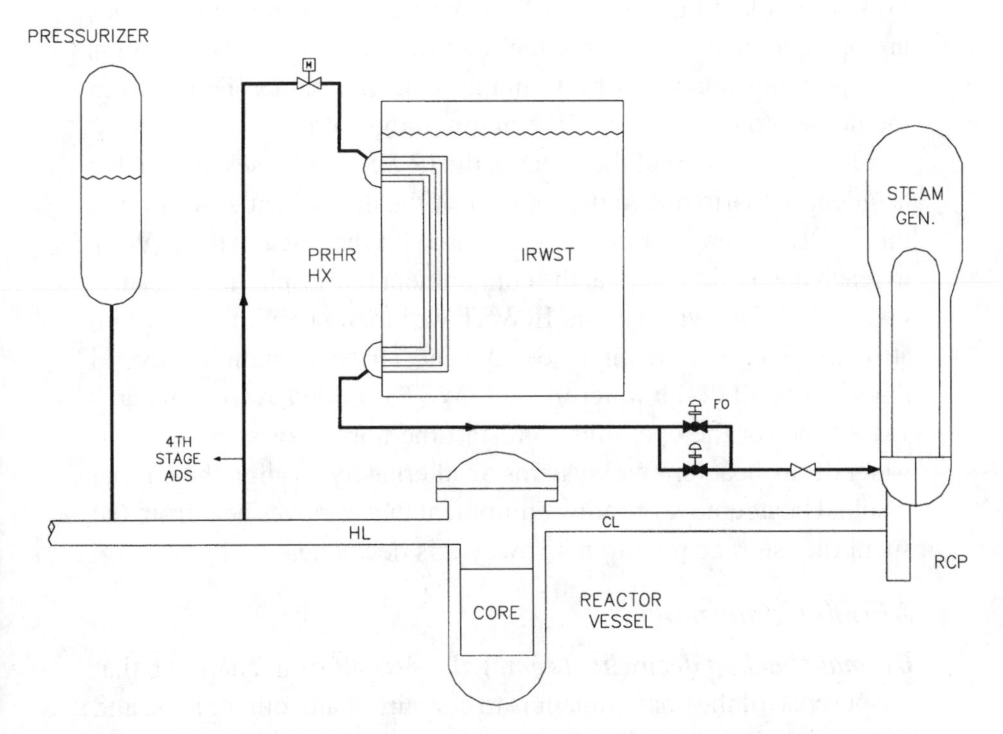

Figure 7.6 AP600 passive residual heat removal system (PRHR). *Courtesy of* WESTINGHOUSE ELECTRIC CORPORATION.

side is connected to the hot leg. The PRHR is separated from the primary system by air-operated valves that "fail open" upon loss of electricity, meaning just as it sounds. The valves are shut against the force of air pressure by electric power during normal operation, but they will be opened by this air pressure in the event that electric power is lost.

Now let us suppose that we lose the normal feedwater decay heat removal systems. Let us also suppose that we lose all electric power and, hence, the primary pumps. The air-operated valves connected to the PRHR are automatically opened since they fail open upon the loss of electric power. Since the PRHR and IRWST are at an elevation higher than the primary coolant system, hot water from the primary system flows by natural convection from the hot leg to the PRHR, where the decay heat is transferred to the water in the IRWST, and the cooled water returns by natural convection to the primary system. In the event that electric power is not lost and the primary coolant pumps are still operating, the air-operated valves are opened on the signal that the feedwater heat removal system is not operating, and water flows in the same direction as before from the hot leg to the suction of the pump in the cold leg.

The temperature of the water in the IRWST tank rises during this accident scenario due to the transfer of the decay heat to it through the PRHR. Several hours will be required for the water in the IRWST to reach the boiling point at the containment (atmospheric) pressure, i.e., 100°C. The water in the IRWST will then begin to boil, while still effectively removing the decay heat. This can go on for several days before all of the water in the IRWST is boiled away. This provides time for the operators to restart the normal or standby feedwater decay heat removal systems or, alternately, to align the normal residual heat removal cooling equipment that removes heat from the spent fuel storage pool to take away this decay heat.

AP600 Containment

Ultimate backup decay heat removal. Recall from chapter 6 that the purpose of the containment is to contain steam, other gases, and radioactive fission products so that radioactive materials will not be released to the environment during an accident. There is still the need to remove the decay heat from the containment even if the core is kept covered with water. In the accident descriptions above, all we did was get the decay heat from the fuel to the large backup source of water. And remember, for these accidents we started with the assumption that the normal decay heat removal systems are not working. Thus, we need an alternate, or backup, decay heat sink for the containment. This is where the ingenious new containment design of the AP600 comes in, as illustrated in figure 7.4.

The decay heat causes some of the water covering the primary system eventually to evaporate. The steam so generated will condense on the walls of the containment as long as the walls of the containment are cooled. Thus keeping the walls of the containment cooled is an ultimate way of removing the decay heat from the containment.

Removing the heat from the walls of the containment is accomplished in the AP600 with a revolutionary new passive containment cooling system (PCS). The actual containment is the steel vessel, or shell, inside the containment building, seen in figure 7.4. The building is concrete with air inlets near the top. There is an additional steel cylinder, or baffle, between the concrete building and the steel containment vessel. The baffle is open at the bottom. Air can flow downward between the concrete building and the baffle, make a u-turn at the bottom of the baffle, and flow upward between the baffle and the containment vessel. There is a large opening at the top of the containment building through which the air can exit. The air flows by ***natural convection***, without the need for electrically driven fans or blowers. Air is heated as it passes by the hot containment shell so that it rises. Cold air from the outdoors flows downward outside the baffle.

This air flow by itself is insufficient to remove all of the decay heat during the first few days of the accidents discussed here. Therefore, during the first few days of an accident, the containment vessel is cooled by water. The source of this water is a large, 360 000 gallon PCS water storage tank at the top of the concrete shield building, outside and above the containment vessel, as also illustrated in figure 7.4. If signaled by a rise in containment pressure or temperature, valves to this water storage tank are automatically opened to allow the water to flow out of the tank. The water flows by gravity over the top of the containment. Water is evaporated as it covers the hot containment, thus cooling it. The vaporized steam flows out through the opening in the top of the containment building with the natural convecting air flow. Thus the water, together with the flowing air, provides the ultimate heat sink for the containment.

The tank contains a three-day supply of water. Operator action is required before the end of the three-day period to refill the tanks. Even if the tanks are not refilled, however, the containment will still not fail since flowing air is also removing heat from it, and after three days the fission product heat source has decayed substantially. If the tank is not refilled, the pressure inside the containment would rise slightly above the design pressure of the containment vessel, but this pressure increase would be insufficient to cause containment failure.

Removal of fission products. In order to reduce off-site doses from fission product aerosols, such as radiative iodine and cesium (which might be in the containment) the concentrations of these fission products must be reduced by a washing action that removes the aerosols. An active containment system is used in conventional reactors, as described in chapter 6; in the AP600, however, the washing action is provided by the steam condensate as it returns to the containment sump. Active sprays are not required.

SBWR, AN ADVANCED PASSIVE BWR DESIGN

The simplified boiling water reactor is being designed by General Electric. Similar to the AP600, its power level is 600 MWe—about half the size of most BWRs now in commercial operation. It has advantages similar to those of the AP600 in that it can be built in a short time (three years), many of its systems can be manufactured and assembled at a central location and transported directly to the plant site for installation, and its design is greatly simplified over earlier BWRs. General Electric has recently suspended development of the SBWR due to market forces, but the passive safety design is of interest here for what may come in the more distant future.

The most remarkable feature of the SBWR is that no pumps are required to recycle the water inside the reactor pressure vessel. Go back to figures 2.14 and 2.15 to review how water is recirculated in a conventional BWR by means of recirculating pumps. It is these recirculating pumps that are entirely eliminated from the SBWR design. Thus, there are no recirculation pipes to break, and it is the rupture of these pipes that constitutes the main accident against which current BWRs must guard. (It should be noted that the evolutionary ABWR design also eliminated this recirculation piping—as Swedish BWRs have for some time. The ABWR and the Swedish BWRs still use recirculation pumps, but they are internal to the reactor vessel, thus eliminating the external recirculation piping.)

How does the SBWR manage this remarkable simplification? The trick is our old friend described above—natural convection. The SBWR is illustrated in figure 7.7. The reactor vessel is very tall. Water in the reactor is boiled and rises above the core beneath the steam separators. This two-phase mixture of water and steam has a very low average density; thus, we see a tall region of low-density fluid, called the chimney. On the outside of the steam separators is the downcomer region where saturated water and incoming feedwater mix and flow down on the outside of the core, outside the core barrel. The colder, higher density water flows downward outside; the hotter, two-phase low-density water/steam mixture flows upward on the inside. This creates the large pressure differential that serves as the driving force for the natural convection flow of the recycle water back down and up through the core. No pumps are needed. The only pumps in the main water/steam system of the SBWR are the feedwater pumps outside the containment that are part of the Rankine cycle/turbine-condenser system (as described in chapter 5). These pumps are not needed for safety, however, since natural convection keeps water flowing through the core in any emergency, together with the also remarkable passive heat removal systems that we will consider next.

It is interesting from an historical point of view to note that the first BWR, the Experimental Boiling Water Reactor built at Argonne National Laboratory, used natural convection for flow through the core. Later, forced recirculation pumps were added by General Electric to

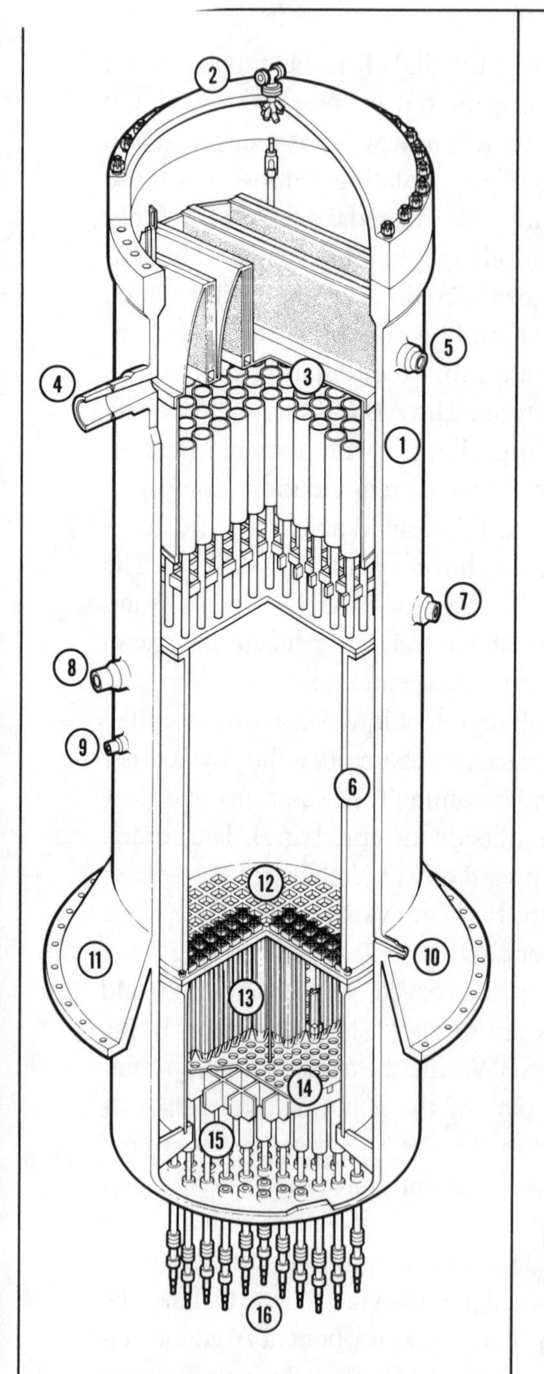

Simplified Boiling Water Reactor Assembly

1 Reactor Pressure Vessel
2 RPV Top Head
3 Integral Dryer-Separator Assembly
4 Main Steam Line Nozzle
5 Depressurization Valve Nozzle
6 Chimney
7 Feedwater Inlet Nozzle
8 Reactor Water Cleanup/Shutdown Cooling Suction Nozzle
9 Isolation Condenser Return Nozzle
10 Gravity-Driven Cooling System Inlet Nozzle
11 RPV Support Skirt
12 Core Top Guide Plate
13 Fuel Assemblies
14 Core Plate
15 Control Rod Guide Tubes
16 Fine Motion Control Rod Drives

GE Nuclear Energy

Figure 7.7 General Electric's SBWR, Simplified Boiling Water Reactor. COURTESY OF GENERAL ELECTRIC COMPANY.

provide increased flow as the power density in commercial plants was increased. The power density in GE's SBWR has been reduced back to a lower level to allow the use of natural convection again.

SBWR Emergency Core and Containment Cooling

What happens if a steam line or a feedwater inlet pipe ruptures so that coolant is lost (loss of coolant accident, or LOCA) and all on and off-site electricity is lost? How is the core cooled, and how is the decay heat removed from the SBWR containment in this situation?

First, isolation valves isolate the reactor from the steam system just as in conventional BWRs. Then the steam is condensed in a pressure suppression pool also as in conventional BWRs, but here the similarities end. After the reactor is depressurized, the core is flooded with water from a gravity driven core cooling system (GDCS), shown in figure 7.8. This system consists of a large pool of water located above the core so that water can flow through pipes into the core by gravity; hence, no pumps are needed. The pressure suppression pool is also located above the reactor core, and this water also can be added to the reactor vessel. Adding water from the GDCS and the suppression pool will flood the entire containment (i.e., the drywell) to a height of at least one metre above the top of the core. Consequently, the core will remain adequately cooled for an indefinite period following a LOCA.

Long-term heat removal and containment cooling is achieved by means of an isolation condenser (IC). Following blowdown, after the pressure in the drywell containment becomes atmospheric, steam from the containment flows by natural convection through tubes in the isolation condenser, which is located in a large isolation condenser pool of water located above the GDCS, as shown in figure 7.8. There the steam condenses by transferring heat to the pool, and the condensate flows back to the reactor. Steam is generated in the IC pool as a result of condensing the steam from the containment; this steam is vented to the atmosphere. Thus, under emergency conditions, decay heat is released to the atmosphere by boil-off from the isolation pool. There is sufficient water in the IC pool to handle three days of decay heat removal. After three days, the operator must

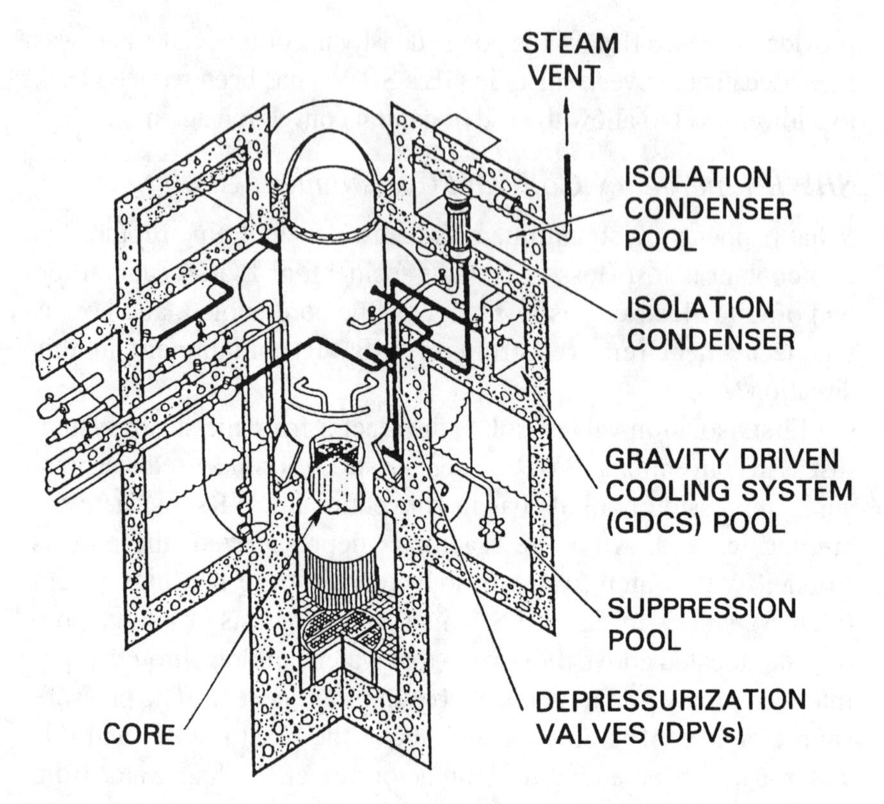

Figure 7.8 SBWR passive emergency cooling systems, showing components that keep the core covered with water in a loss of coolant accident and that take away the decay heat. COURTESY OF GENERAL ELECTRIC COMPANY.

add water to the pool to continue this passive containment cooling indefinitely.

For a small pipe break or leak in the primary system, the reactor is isolated from the steam system, and the reactor is depressurized by means of six depressurization valves located on steam pipes in the upper drywell. After depressurization, the GDCS can provide water to the vessel by gravity flow as required. Again decay heat can be removed by the isolation condenser.

For non-LOCA events in a conventional BWR in which the reactor is isolated from the turbine condenser, pressure must be reduced by venting steam through relief valves into the suppression pool. For most such events in the SBWR, however, pressure can be controlled automatically by means of the isolation condenser without removing water from the reactor vessel. Steam is diverted to the isolation condenser, and its heat is released to the IC pool. Again flow is by natural convection, so no pumps are used. If needed, increased pressure can be relieved by means of four conventional pneumatic/spring-actuated safety relief valves. Present BWRs have about twenty such relief valves. This reduction in number for the SBWR is possible because of the low rate of pressure increase for the SBWR in accident situations.

Valves that separate the GDCS, isolation condenser, and the suppression pool from the vessel must open and shut to allow water to flow into and out of the vessel and containment, depending on the accident scenario. For example, in a pipe break these valves are "explosively" operated valves that function automatically when steam pressure is sensed in the drywell, without operator action. The valves operate on direct current from a battery. This means that they are operated by stored energy rather than by an active source of alternating current so that they do not require either off-site power or an on-site diesel generator. Like the AP600, the SBWR has diesel backup ac power, but the diesels are not "safety related" since they are not required for emergency core or containment cooling.

GAS TURBINE-MODULAR HELIUM REACTOR (GT-MHR)

The General Atomic Company in La Jolla, California, is designing a gas-cooled reactor that has three revolutionary features. It is modular, it uses a gas turbine, and it is automatically safe. It's called the Gas Turbine-Modular Helium Reactor (GT-MHR).

Gas has been used as a reactor coolant from the beginning, for plutonium production reactors during the Second World War and for the first commercial power reactors (in the United Kingdom and France). Modern gas-cooled reactors use the inert gas, helium. Gas cooling has the advantage that the helium coolant can be operated at much higher temperatures than the water in water-cooled reactors. This leads to higher thermal efficiencies for the power cycle, which, you will recall from chapter 5, means that a higher fraction of the heat can be converted into electricity. In the United States, General Atomic built and operated a commercial helium-cooled power reactor at Fort St. Vrain, Colorado. That reactor is decommissioned now, but the fuel used for this reactor is essentially the same as the fuel that will be used in the GT-MHR, and the fuel in Fort St. Vrain worked well. Another U.S. company, Adams Atomic Engines, is also developing a gas-cooled reactor; more on their design appears in the discussion of ***pebble bed reactors*** at the end of this section on gas-cooled reactors.

General Atomic had designed a modular gas-cooled reactor that was to use a steam turbine, but, in a dramatic switch, they recently changed the design to a gas turbine. All design parameters are not set at the time of this writing, but much is still known about the new design. The electrical power rating of each module for the gas-turbine design is between 250 and 300 MWe. There may be several modules in each power station, with one reactor for each module. The helium coolant gas is maintained at high pressure [69 atmospheres (1017 psi, 7 MPa)] in the reactor in order to provide satisfactory heat transfer to take the heat away from the fuel.

This reactor has so many dramatic features that it's hard to know where to begin. I'll choose the gas turbine feature since that is what's so new to the nuclear industry. Gas turbines are being used for natural gas electric power plants in great numbers nowadays. This development resulted from the development of the airplane jet engine; a gas turbine in an electric power plant is simply a very large jet engine. Modern engineering and material development have made this gas turbine system economically competitive. The General Atomic people decided the time had come to use this technology for their high-temperature helium-cooled reactor.

The power cycle is illustrated in its simplest form in figure 7.9. This cycle replaces the BWR steam cycle shown back in figure 5.10. The helium gas leaving the reactor enters the gas turbine, which drives the generator to make electricity. No need to boil any water. The gas from the turbine is cooled and sent to the compressor. Here

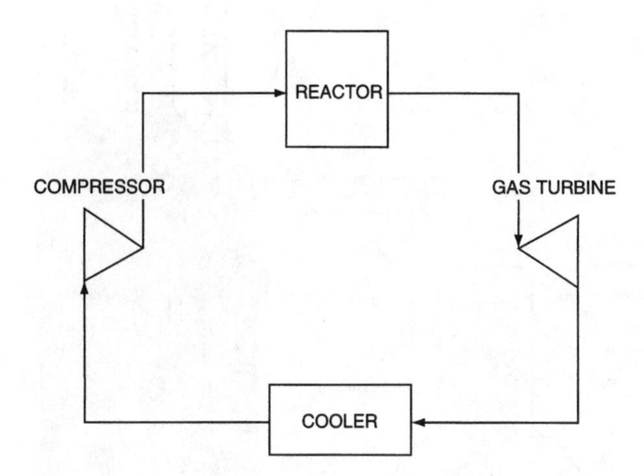

Figure 7.9 Simplified gas turbine power cycle.

it is compressed back up to the reactor pressure and returned to the reactor. Like the steam cycle, this gas power cycle has a name; it is the ***Brayton cycle***.

The more complete arrangement for the GT-MHR is shown in figure 7.10. The reactor is in the vessel on the right. Note that the helium flows downward through the core. The outlet helium temperature is 850°C (1562°F), which is much higher than that of the water in LWRs. Back-up helium circulators are located below the core. The power production vessel on the left houses the turbine, generator, compressor, and coolers. You will notice that there are more components than I showed in figure 7.9; the precooler, intercooler, and recuperator are refinements that make the system more efficient. The thermal cycle efficiency is 47%, which compares to 33% for light water reactors. The coolers are helium-to-water heat exchangers that replace the condenser in the steam power cycle; hence, cooling towers, lakes, or rivers are still needed as ultimate heat sinks. To me it's amazing to think of all this power equipment in one vessel after getting used to the large pieces of equipment in a steam plant, which require an entire building. For earthquake

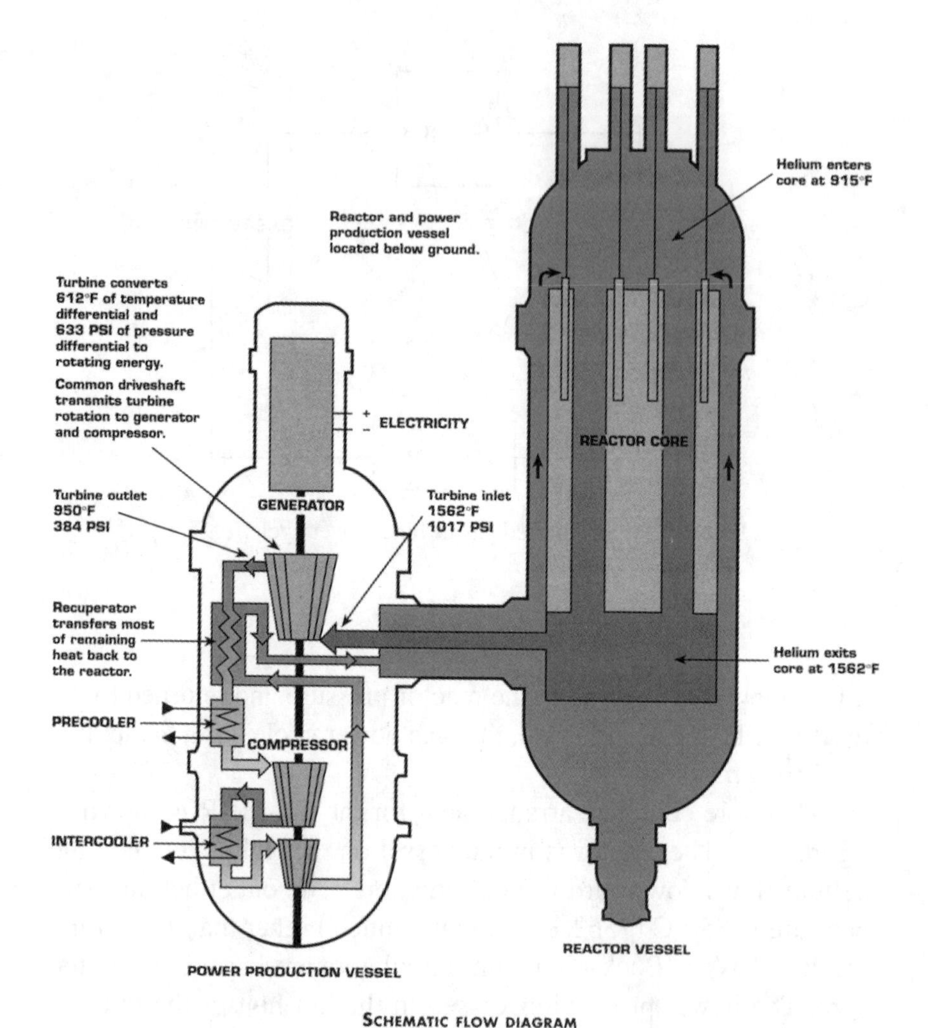

Figure 7.10 Schematic flow diagram for the GT-MHR. COURTESY OF GENERAL ATOMICS.

protection, the reactor and power production vessels are located underground in a silo.

The fuel in a GT-MHR is entirely different from the other reactors that have been discussed so far. The fuel is in the form of tiny, half-millimetre diameter particles, as illustrated in figure 7.11. The center of the fuel particle is uranium oxycarbide, meaning that 70%

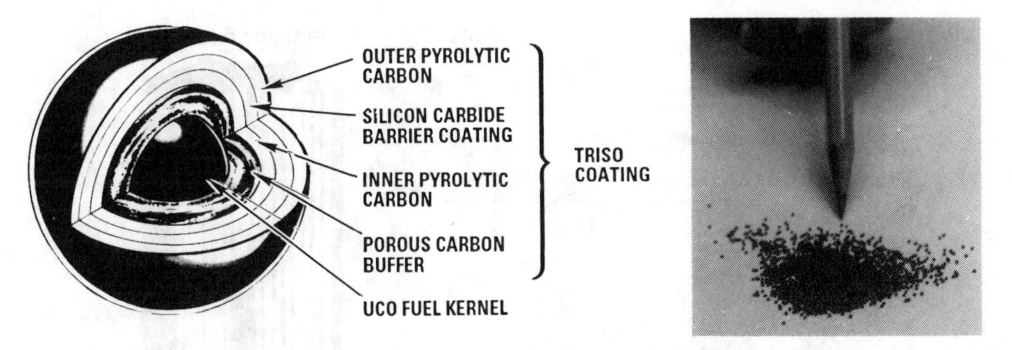

Figure 7.11 Fuel particle for the Gas Turbine-Modular Helium Reactor. COURTESY OF GENERAL ATOMICS.

of it is UO_2 and 30% is UC_2. The fuel is encased in several layers of carbon bearing material. The first is a porous carbon that can absorb the impact of fission products and fission product gases such as xenon and krypton. The next layer is a "pyrolitic carbon." The third layer is silicon carbide, a material that has strength and can contain the pressure built up by fission product gas. The outer layer is another layer of pyrolitic carbon. This fuel has a melting point well above 2000°C. It can contain all of the fission products as long as the fuel temperature remains below 1600°C, and most of the fission products even if the fuel temperature were to rise to 2000°C. During normal operation, the maximum fuel temperature is 1100°C.

The fuel is embedded in a graphite matrix in the form of rods. These rods are placed in hexagonal graphite blocks, as shown in figure 7.12, and the blocks are stacked in columns. Graphite is a form of carbon. Holes are drilled in the graphite to allow for the flow of the helium coolant. Thus, the heat passes from the fuel particles to the graphite fuel rods, then to the graphite, and finally to the helium. The graphite serves as both the structural material and the neutron moderator. The core layout is illustrated in figure 7.13. The core is annular (like a very thick cylindrical pipe), with a graphite reflector both inside and outside the core. There are 102 graphite/fuel columns, 48 control rods, and 18 backup control channels into which

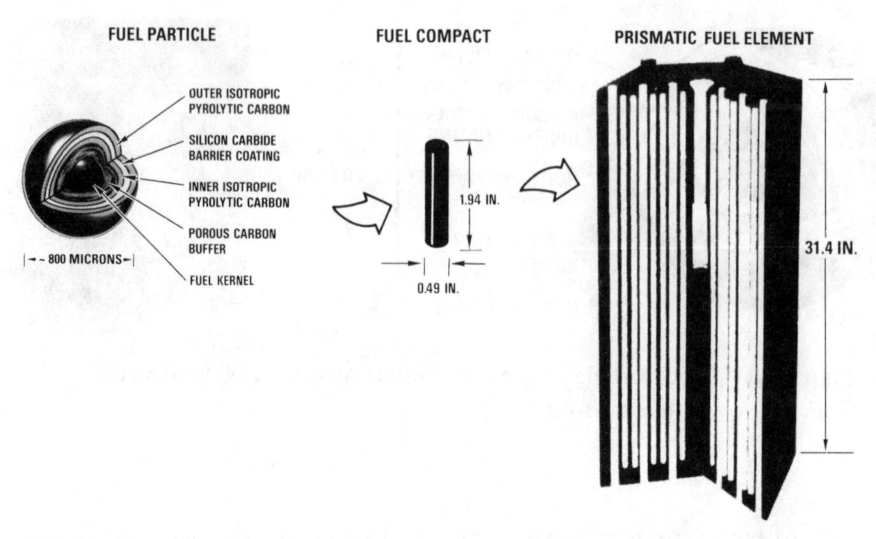

Figure 7.12 Graphite fuel blocks for the GT-MHR. COURTESY OF GENERAL ATOMICS.

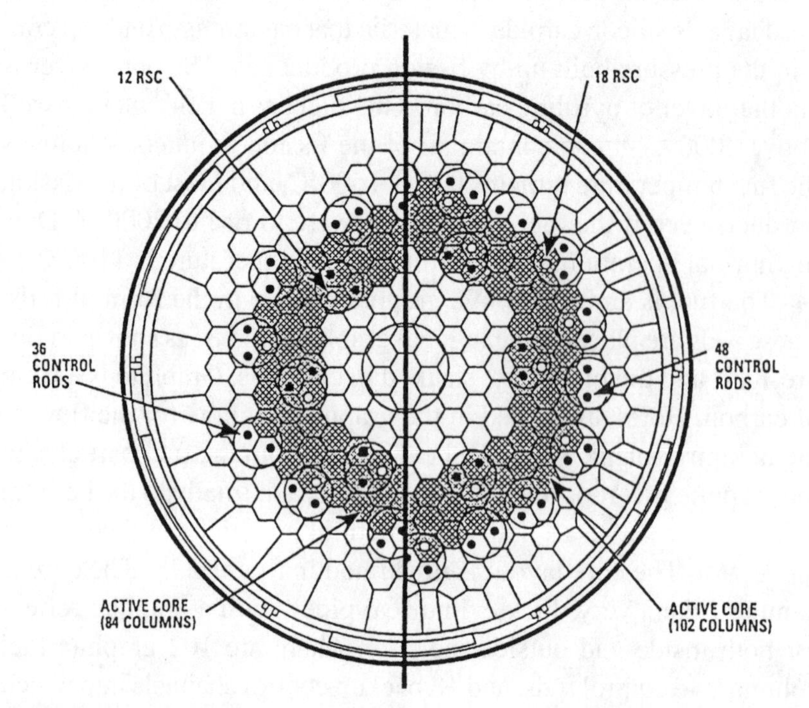

Figure 7.13 GT-MHR Core Design. COURTESY OF GENERAL ATOMICS.

marble-size boron balls can fall in the event that distortion of the core geometry prevents insertion of the control rods.

Now come the dramatic passive safety features of this reactor design. There are two. First, the fuel particle is the "containment." In no accident is it possible for the fuel temperature to rise above about 1600°C. Therefore, there is no chance that fission products can escape from the fuel to endanger the public and no chance of a meltdown. Thus there is no credible reason to provide an additional containment beyond the fuel itself. (Society may demand an additional containment—it is too early to tell—but from any technical point of view there is no need whatever. However, as we also know, society is not always guided by rational decisions.)

The second feature is the reason why there is no chance that the fuel can ever rise over about 1600°C; this is an ingenious feature called the reactor cavity cooling system (RCCS), illustrated in figure 7.14. You recall that decay heat must be removed from a reactor after the reactor is shut down. In normal operation this is accomplished by the coolers in the power production system, which require electricity for their operation. However, in an accident situation such as the complete loss of electric power, decay heat can be removed by the RCCS.

Now let's examine how the RCCS works. The RCCS is an air-flow system consisting of four intake and exhaust structures through which the air passes and an air-flow path into the reactor cavity and around the reactor vessel. An annular baffle separates the cold air entering the system from heated air flowing near the reactor vessel. Outside air, which is "cold," flows downward between the baffle and the reactor cavity wall. The air then makes a u-turn and flows upward between the reactor vessel and the baffle. The reactor vessel is hot since heat is being transferred through the vessel from the graphite and fuel in the core. The rising air is heated by heat transfer from the reactor vessel. Since the air flowing past the reactor vessel is hotter than the colder outdoor air flowing downward outside the baffle, the air flows by natural convection. Again, no fans are used to make the air flow; only natural forces that cannot fail are used. Natural convection provides the ultimate heat sink just as was

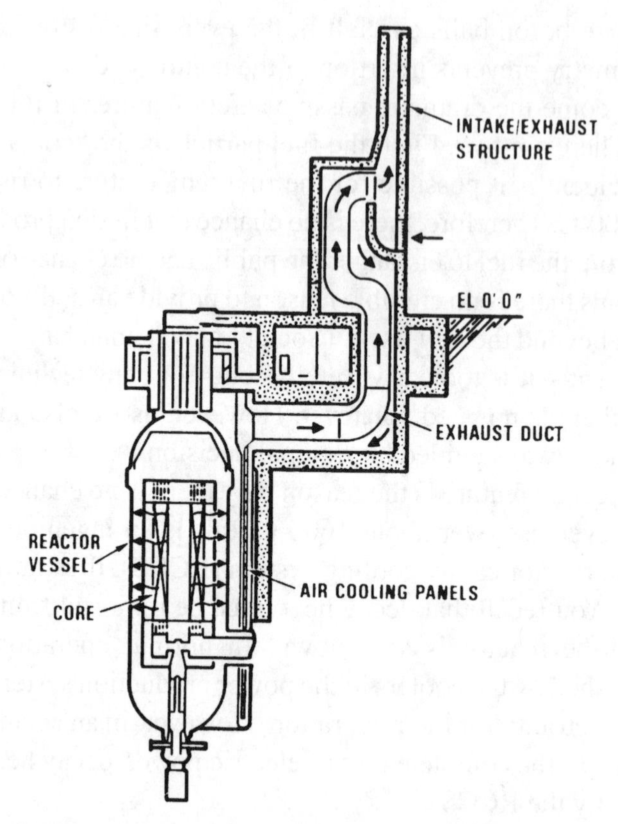

Figure 7.14 GT-MHR reactor cavity cooling system (RCCS). COURTESY OF GENERAL ATOMICS.

the case with the AP600 and the SBWR. Neither electricity nor operator action is needed; the system is entirely automatic. The air in the RCCS would be flowing even during normal operation; thus, no mechanical valves or louvres need to open for the system to work.

Normally the reactor would scram in an accident even if all electricity were lost. However, even if the reactor did not scram, the criticality factor would be reduced such that the power would drop to such a low level by natural processes accompanying a rise in fuel and graphite temperatures that the RCCS would remove all of the decay heat. The fuel never exceeds 1600°C in such an event, so no fission products are released. As far as the fuel temperature and the protection of the public are concerned, it does not matter whether the helium is still in the reactor at high pressure or whether the system has depressurized and the helium has been lost. The decay heat is still taken away by the air, and the fuel temperature still remains low. The RCCS can take away the decay heat forever, i.e., as long as it takes to turn the electricity back on and make the compressor or backup circulator and the normal decay heat removal systems work again.

Now you can raise some questions. What if the RCCS gets plugged? Almost impossible, but not absolutely impossible. Then the decay heat is transferred to the ground around the reactor cavity with the fuel still not rising above 1600°C. A problem arises with the reactor vessel in this case; the vessel temperature gets so hot that it may not be possible to use it again, but this has no bearing on whether the public is protected in the accident. You might wonder whether the graphite can catch on fire. There are two answers. The first is that it can burn only if air can get to the graphite in a chimney effect, i.e., if there is an opening for the air both above and below the core so that the air can flow through the graphite in the core by a natural convection process. Experiments have proved this. The probability of this is rather incredible, though not impossible. The second answer is that if the graphite were to burn, the fuel temperature would still not rise above 1600°C, and, again, none of the fission products would be released to the environment.

General Atomic's analysts predict that the GT-MHR can produce electricity at a price lower than light water reactors, coal, and natural gas. It appears that it's next to impossible for this reactor to harm the public in any accident, but more research is needed to prove this. It will be necessary to build a prototype plant to confirm both safety and economics. Thus, it will be some time before we will know the place of the GT-MHR in the mix of electricity generation.

Pebble Bed Reactor

There is a variation of the high-temperature gas-cooled reactor called the ***pebble bed reactor***. In this design the fuel is still in the form of small carbon-coated particles, but they are put into a graphite matrix in the form of balls the size of pool cue balls, or tennis balls. These balls, or pebbles (though rather large pebbles), are loaded into a vessel so that this becomes the reactor core. Helium then flows down between the balls. New balls are loaded at the top, and burned balls are removed from the bottom, so that refueling occurs without even shutting the reactor down.

This type of reactor has been built and operated in Germany for a number of years, at a power level of about 300 MWe. A dramatic picture of the German's pebble bed reactor is shown in figure 7.15.

A Florida company, Adams Atomic Engines, is developing a gas-cooled reactor for small power plant applications, such as merchant

Figure 7.15 Pebble bed reactor, an alternate design for a high-temperature gas-cooled reactor. COURTESY OF GESELLSCHAFT FÜR HOCHTEMPERATURREAKTOREN.

ship propulsion and electricity generation for remote areas. They have chosen the pebble bed reactor for their design.

ADVANCED LIQUID METAL REACTOR (ALMR)

The liquid metal reactor (LMR) differs from light water reactors and gas-cooled reactors in three major ways. First, the reactor is cooled by the liquid metal sodium. Second, the neutron spectrum is fast instead of thermal; thus, it is called a fast reactor. Third, the LMR is a breeder reactor.

A breeder reactor generates more fuel than it consumes. This may sound too good to be true, like a perpetual motion machine. In this case, however, it is real, though we need to get our semantics straight. The fuel for a fast reactor is plutonium. In a breeder reactor, as discussed in chapter 5 (page 145), plutonium is produced from uranium from neutron capture in uranium-238. A certain amount of plutonium is consumed during the operation of an LMR. While this plutonium is being consumed, however, new plutonium is continually being produced from uranium, which is also in the fuel. And, in fact, more plutonium is produced than is consumed. Hence, the name breeder. The extra plutonium can be removed and used to start up an entirely new reactor.

The implications of the breeder reactor are important. While I said the fuel was plutonium (since this is the main element that is fissioning), one can consider the fundamental fuel for the breeder to be uranium-238 since this material is used to produce the plutonium. There is enough uranium-238 in the world to last forever. Thus, the LMR represents an inexhaustible source of energy. Using uranium in a light water reactor (LWR) requires the use of U-235, which is not abundant in nature. The use of U-235 implies the use of low-cost uranium from high-grade ores, which are limited in supply. Thus, the economical use of the LWR is limited to some time well into the twenty-first century. For the breeder, however, the ultimate fuel is U-238, and U-238 is so abundant that it will be economical forever. It would be economical to obtain U-238 from seawater, for example, for use in a breeder reactor. The amount of uranium in the oceans is enormous, and this amount is constantly growing as a result of erosion of land by rivers. Thus, the supply of U-238 is virtually limitless. Before we worry about having to obtain uranium from the ocean, however, it should be noted that there is already enough depleted U-238 stored in the United States at U-235 separation plants to generate all of the United States' electricity for literally hundreds of years. By the time the world runs out of low-cost natural uranium, there will be enough depleted U-238 stored at almost no cost to produce the entire world's electricity for hundreds of years.

Liquid metal fast breeder reactors may not be economically competitive for a number of years. However, I believe it is reasonable to make the following rather dramatic speculation. For the long term, i.e., beginning sometime in the late twenty-first or early twenty-second century and probably lasting for many years beyond that (forever, unless something more economical comes along), the fast breeder reactor may become the world's primary source of electricity.

Liquid metal reactors have been operating for over forty years. The first electricity ever generated by a nuclear reactor came from the "Experimental Breeder Reactor I" in 1951, in Idaho. There is a full-sized LMR power plant, with a 1200 MWe power rating, called Superphoenix, currently operating in France. A 600 MWe LMR is operating in the USSR. Smaller LMRs in the 250 to 350 MWe range are operating in Japan, France, and the USSR, and one has operated in the United Kingdom. A 400 MWt LMR test reactor called the Fast Flux Test Facility operated for two decades in the United States, in Washington state. These reactors all use a common fuel—mixed plutonium-uranium oxide. This is basically the same ceramic used in light water reactors except that the fraction of plutonium is approximately 20% or more and the uranium is nearly all U-238 in the LMR, while the U-235 fraction in the LWR is only 3%, with a small amount of plutonium built up during operation. These reactors are all breeders, but electric power from them costs more than from LWRs since they are expensive to build, and this cost disadvantage will remain as long as low-cost uranium is available for use in LWRs.

Two features of an LMR that distinguish it from other designs are worth mentioning. First, the LMR is a fast reactor, as discussed in chapter 5. This means that fissions are caused by fast, high-energy neutrons rather than thermal neutrons. The second feature is that it operates at atmospheric pressure. Sodium boils at 883°C (1620°F). The sodium temperature during normal operation is several hundred degrees below this value so that there is no need to pressurize the coolant system to prevent boiling.

A new LMR design, called the ***Advanced LMR***, has been developed in the United States, though not yet built. This design will be described here because of its features superior to previously built LMRs. The particular features that make the ALMR exciting are its passive safety characteristics. The design is almost completely foolproof against fuel melting. Its safety depends entirely—100%—on natural forces and phenomena, with no need for operation of any mechanical equipment. Mechanical components such as control rods do control the reactor during normal operation, but they are not required to function to prevent damage to the reactor.

A variation of the ALMR is called the ***Integral Fast Reactor*** (IFR) because its reprocessing and fuel fabrication plants are located on the same site as the power plant. An especially attractive feature of the IFR reprocessing and fabrication facility is its resistance to the diversion of plutonium for use in nuclear weapons. Even though the fuel cycle uses plutonium, the type of reprocessing used ensures that the plutonium is always so contaminated with other materials that it cannot be used to make a nuclear explosive.

Further development of the ALMR was recently stopped in the United States since it will not be needed for such a long time [and since a political assessment (though misguided, in my view) that the use of plutonium in the ALMR/IFR might lead to nuclear weapons proliferation is currently in vogue]. However, I will describe the reactor here, especially its remarkable passive safety features, since I expect we will one day return to this, or a similar, concept.

As stated earlier in this chapter, the ALMR is a modular reactor. An entire ALMR plant generates 1395 MWe of electricity. The plant consists of three modules, or power blocks, each of which generates 465 MWe. Furthermore, each module consists of three reactors, so that each reactor is generating only 155 MWe of electric power. Thus there are nine reactors in the entire plant. The thermal rating of each of the nine reactors is 471 MWt. (Recall that the distinction between MWe and MWt was described at the beginning of chapter 5.) A utility or power generating company can order one, two, or three modules at a time.

Another characteristic of the ALMR that distinguishes it from the other LMR power reactors is its fuel. The ALMR uses metallic fuel instead of oxide fuel. The fuel is 10% (by weight) zirconium, about 27% plutonium, and the remainder depleted uranium.

The ALMR plant, showing the three power blocks, is illustrated in figure 7.16. The reactor steam supply system is shown in figure 7.17. An LMR differs from an LWR in that an ***intermediate*** or ***secondary*** coolant system is required for an LMR. The sodium

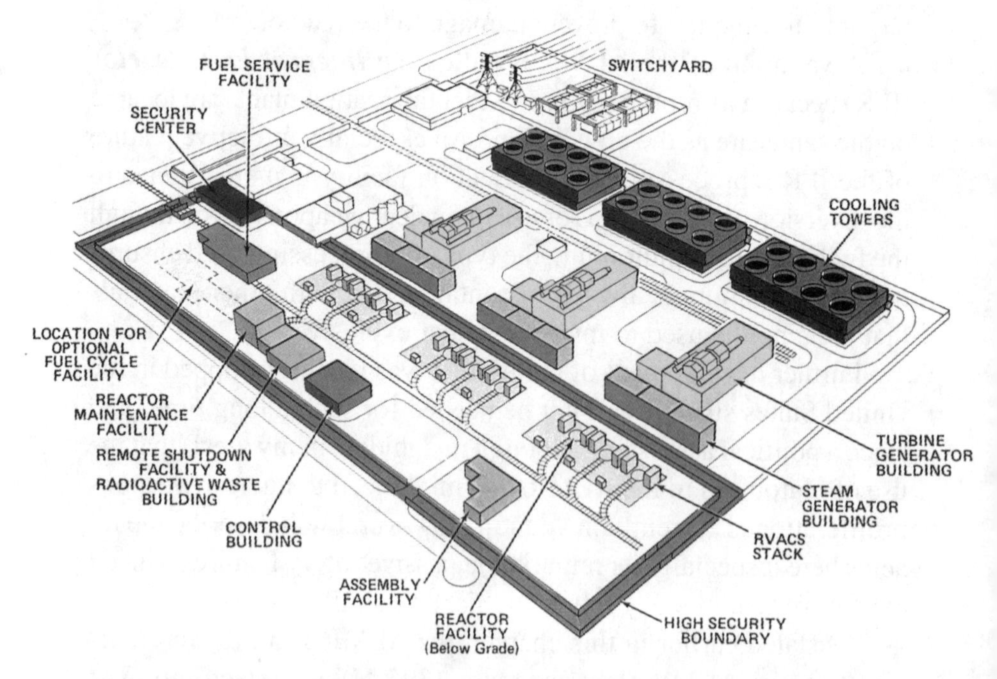

Figure 7.16 Advanced Liquid Metal Reactor power station, showing three power blocks. The optional fuel cycle facility would be on-site reprocessing and fuel fabrication facilities as provided by the Integral Fast Reactor concept. COURTESY OF GENERAL ELECTRIC COMPANY.

coolant from the reactor (the ***primary*** sodium) transfers its heat to the secondary sodium, and this secondary sodium is used to generate the steam in the steam generator. The reason for the need for the intermediate sodium loop is that sodium becomes radioactive as it flows through the reactor. Sodium-23 absorbs a neutron to become sodium-24, which is radioactive, with a half-life of 15 hours. The secondary sodium is isolated from the primary sodium so that the secondary sodium is not radioactive. It is important to have nonradioactive sodium in the steam generator because there is always the chance that a leak could develop there, and some of the sodium could react with the water. Figure 7.17 shows the secondary sodium pipes and the secondary sodium pumps in a typical loop.

The reactor module is shown in figure 7.18. The reactor vessel houses three basic components—the reactor core, the intermediate heat exchangers (IHX), and the primary system pumps. In the

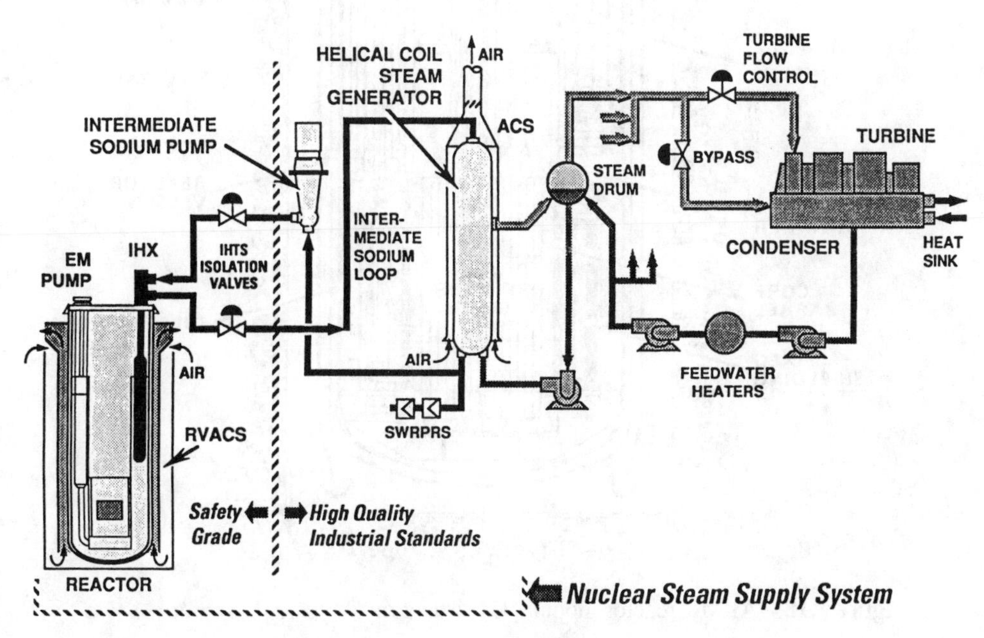

Figure 7.17 ALMR heat transport system and steam cycle. COURTESY OF GENERAL ELECTRIC COMPANY.

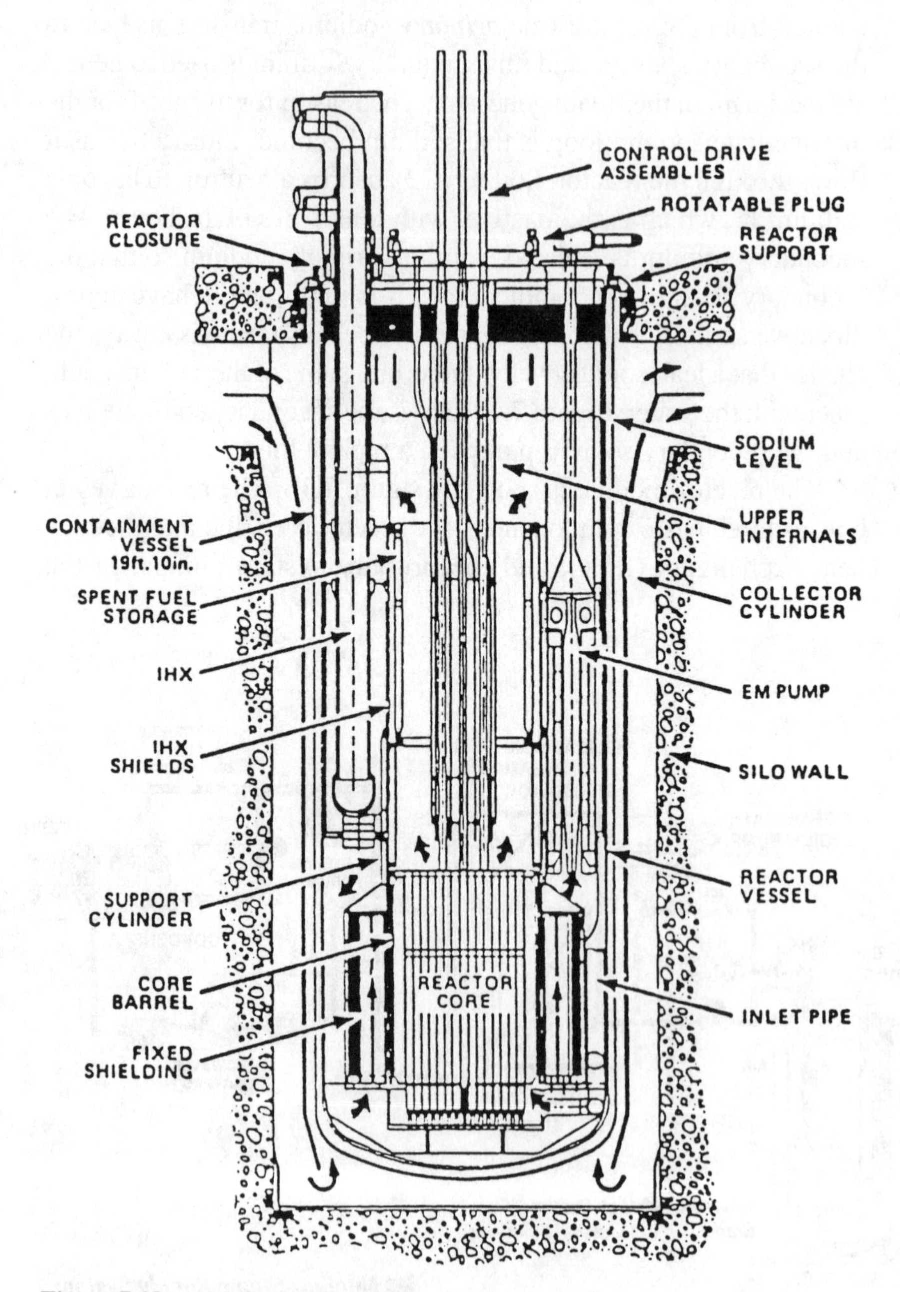

Figure 7.18 ALMR reactor module. COURTESY OF GENERAL ELECTRIC COMPANY.

ALMR, the primary pump is an electromagnetic pump. This is possible since the coolant is a liquid metal; the magnetic field creates a force on the liquid metal (which is not possible for a water or gas coolant) that makes it flow.

Another feature shown in figure 7.18 is the containment, or guard vessel. This is a vessel that surrounds the reactor vessel. If a leak develops in the reactor vessel, the sodium would be caught in the guard vessel so that the core would not be uncovered. This guard vessel also serves as the containment. There is a dome above the vessel cover at the top of the reactor that serves as part of the containment. All of the sodium piping goes through the reactor cover so there are no pipes through the sides of the reactor vessel that could lead to sodium leakage from the vessel. Compressed helium is present in the space between the reactor vessel and the guard vessel so that any potential leak in the guard vessel can be detected before any sodium leaking from the reactor vessel ever enters the guard vessel.

The core design is illustrated in figure 7.19. The core consists of fuel assemblies and internal blanket assemblies. An external blanket surrounds the core, both axially and radially. The fuel is in the form of small-diameter rods (about 6 mm, or 0.25 inch), similar to the fuel in a light water reactor. The core is smaller in an LMR, however, with a height of only about 1.3 m (4.3 ft). The coolant flows upward through the core past the fuel rods, as in an LWR. Control rods use boron as the neutron poison. There are six control rods, with only one actually needed to shut down the reactor. The gas expansion module is a safety feature that will automatically shut down the reactor if the sodium flow stops. The "ultimate shutdown" at the center is a channel into which boron balls can be dumped to shut down the reactor if distortion in the core geometry were to prevent the insertion of all six control rods.

Two outstanding safety features exist in the ALMR that distinguish it from other LMR designs. First, if the coolant flow stops (as would happen in the loss of off-site and on-site electricity) or if the criticality factor increased above unity due to a mechanical failure, ***and simultaneously*** if the control rod scram system completely failed,

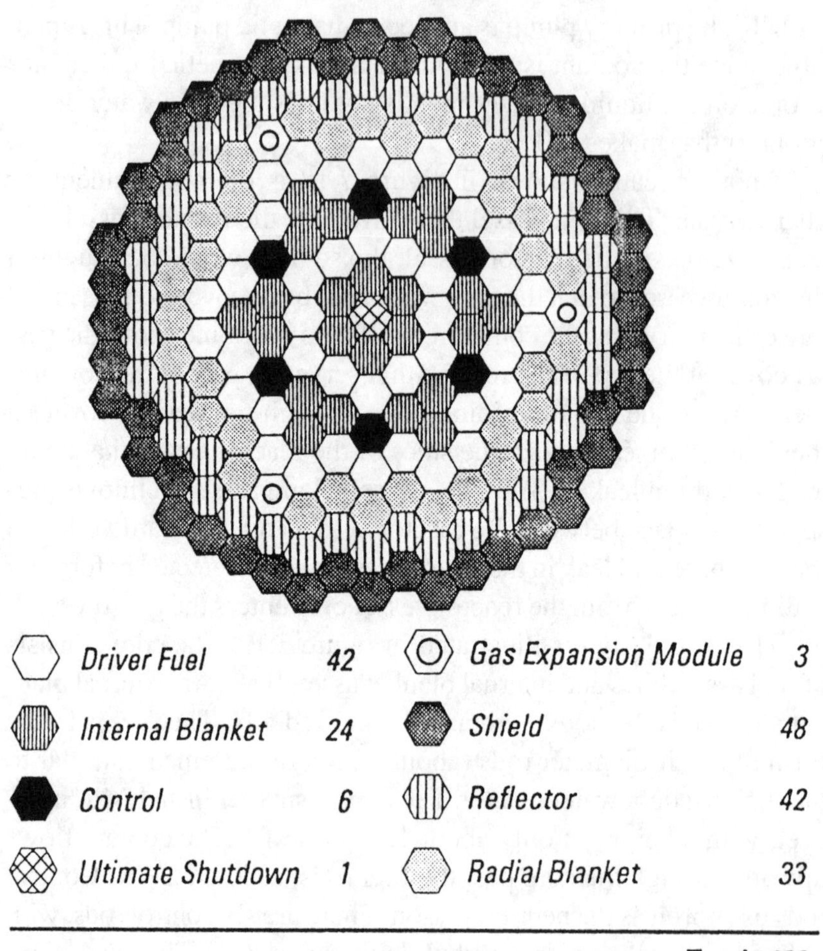

Figure 7.19 ALMR core design. COURTESY OF GENERAL ELECTRIC COMPANY.

the reactor would change in such a way that the chain reaction would either shut down or the power level would be reduced to a low level. The fuel and coolant temperatures would rise in such an accident, and this temperature rise would automatically shut the reactor down, without any operator or mechanical equipment action. The mechanisms that respond to the temperature rise include material expansions and changes in nuclear reaction rates that occur naturally and independently of operator or mechanical actions.

The second feature is natural convection to remove decay heat. Like the passive LWRs and the GT-MHR, the ALMR also takes advantage of this phenomenon for its ultimate heat sink. The ALMR makes use of a system called the reactor vessel auxiliary cooling system (RVACS) to perform this task. The RVACS is illustrated in figure 7.20. Like the GT-MHR's reactor cavity cooling system, it operates entirely by natural convection, it requires absolutely no mechanical action for its operation (it is passive), and it can remove all of the decay heat as long as needed without any operator action.

RVACS is an air-flow system consisting of four large chimneys where the air enters and leaves and an air flow path around the containment, or guard vessel. An annular collector, or baffle, separates

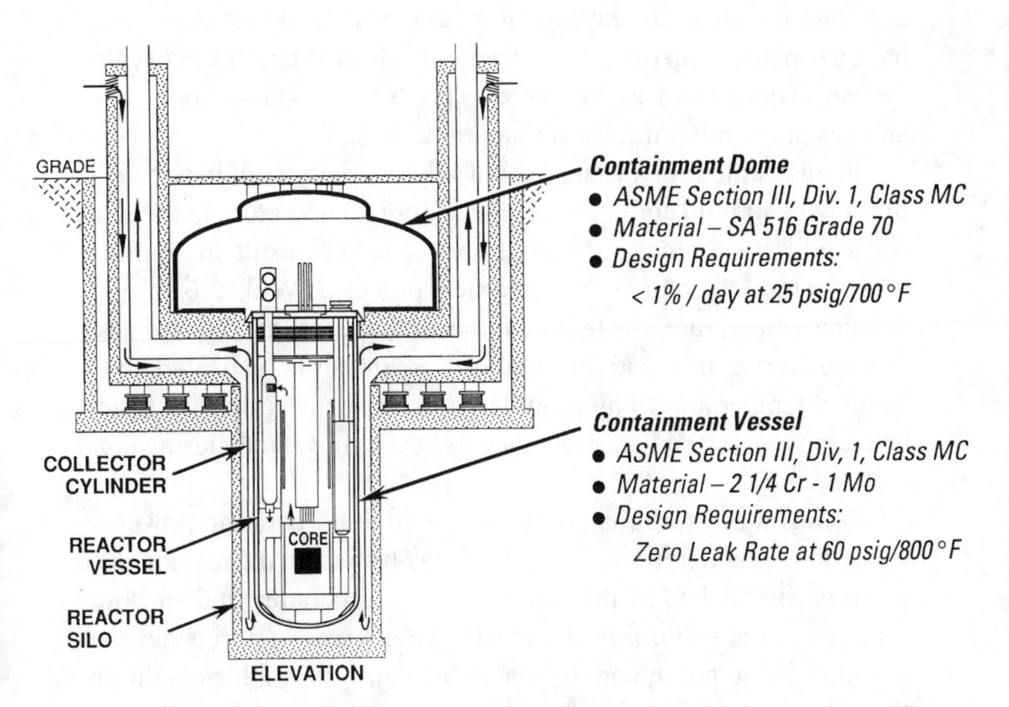

Figure 7.20 ALMR containment and reactor auxiliary cooling system (RVACS). COURTESY OF GENERAL ELECTRIC COMPANY.

the cold air entering the system from heated air flowing adjacent to the guard vessel. The air flows by natural convection. Outside air, which is "cold," flows downward between the collector and the reactor cavity wall. The air then makes a u-turn and flows upward between the guard vessel and the collector. The guard vessel is hot since it is close to the reactor vessel, which is at the temperature of the sodium inside the vessel. The rising air is heated by the guard vessel. Since the air flowing past the guard vessel is hotter than the colder outdoor air flowing downward outside the collector, the air flows by natural convection. Again, no pumps are used to make the air flow; only natural forces that cannot fail are used.

Air flows through the RVACS during normal operation of the reactor. This takes away about 0.2% of the thermal power generated by the reactor; this is just lost to the atmosphere. However, this means that if the RVACS is needed as the ultimate heat sink in an accident, the air is already flowing; no mechanical valves or vents have to open and no operator action is needed to start or modify the system. The system is completely passive, absolutely foolproof, and based entirely on fail-safe natural forces.

In an accident involving loss of all electricity and, hence, loss of primary sodium coolant flow, the sodium temperature in the vessel would rise (which, as stated above, would result in shutting down or greatly reducing the fission process). With this rise in sodium temperature, the temperatures of the reactor and guard vessels would rise also, and this would cause the air to be heated up to a higher temperature. This would in turn cause the air to flow faster and remove heat at a faster rate than its heat removal rate during normal operation.

Removal of more than 0.2% of the normal operating power is needed in order to remove all of the decay heat from the reactor, and this is accomplished by the increase in air temperature and air flow. As a result the sodium in the reactor vessel never approaches the sodium boiling point, and the reactor can just stay in this condition for as long as is needed to fix whatever needs fixing, regardless of how long this takes. Again, no operator or mechanical action is needed; there is no way that the coolant can be lost from the vessel to allow the fuel to melt. Can one conceive of one or more of the four chimneys directing the air flow plugging up? It is difficult to imagine since they are designed to be tornado and earthquake proof, but the system will work even if three of the four chimneys are plugged!

THE PIUS REACTOR

The final advanced reactor that will be described here is the PIUS reactor (which stands for Project Inherent Ultimate Safety). PIUS is a 640 MWe pressurized water reactor, but it has many features that are quite different from a normal PWR. This reactor was invented in Sweden and in some ways was perhaps responsible for initiating the search for design safety based on natural phenomena exemplified by the other advanced reactors described above. The PIUS reactor is the most radical of the advanced reactors. It may be the one with the most inherent safety features. However, it has not yet been built, and a prototype would have to operate in order to convince the industrial reactor community that it is simultaneously economical and as fail-safe as its advocates claim. It is so novel and fascinating in concept that it is worth describing here.

Its inherent safety feature is based on the fact that hot water (which has low density) floats on top of cold water (which is more dense); if hot water is above cold water, the cold water will not flow upward into the hot water. This layering of hot water above cold water has a name; it is called a ***density lock***. Actually most of you have probably experienced this phenomenon yourself when you were swimming in a lake with warm water near the surface, and cold water a few feet below the surface. See, there's a scientific reason for everything. These locks are about one metre deep in the PIUS reactor so the interface between the cold and the hot water does not have to be an absolute plane; there is a gradual transition from cold to hot, but it is sufficient to prevent flow between the cold and hot water.

In the PIUS reactor, the core is placed in an enormous pool of water, as illustrated in figure 7.21. The contents of the vessel, including the reactor core, are maintained at 1300 psi (9 MPa, 90 bars). The core is 2.5 m (8 ft) tall and 3.8 m (12 ft) in diameter. The vessel that contains the pool is 45 m (148 ft) high and 29 m (95 ft) in diameter! This is big—very big. The vessel is made of prestressed concrete, which is a technique to keep the concrete always in tension by means of steel cable embedded in the concrete. This is the only way one can make such a large vessel contain such a high pressure. The inside height of the pool is 34 m (112 ft), and the inside diameter is 13 m (44 ft).

There are two density locks in the coolant system. One is halfway up the tall chimney above the core; the second is below the core. The water in the pool is at low temperature and contains the neutron poison, boric acid. The steam generators and primary pumps are located outside the pool. Water heated in the core flows upward through the center of the tall chimney; cooler water flows downward through the outside annulus and to the bottom of the core. The pressurizer is located at the top of the pool, as shown in figure 7.21. Both the primary system and the entire pool are pressurized to 1300 psi.

During normal operation, water entering the core is hot relative to the water in the pool so this water keeps the cold water in the pool from entering the primary system by means of the density lock below the core. Above the core, the hot water in the primary system keeps the cold water in the pool from entering through the upper density lock. The pumps run at just the right speed so that the hydrostatic head of the pool plus the density difference between the hot and cold water exactly balance the pressure generated by the pump to prevent flow through the density locks.

If the pump speed is increased, the upper lock will be broken, meaning that the pump will draw in more water than needed, and cold water will be sucked in through the upper lock. The lower lock will also be broken, and the excess brought in through the upper lock will go out through the lower lock. The pool water brought in through the upper lock will bring boron in with it, and the reactor will be shut down (or the power will be reduced until the pump speed is reduced).

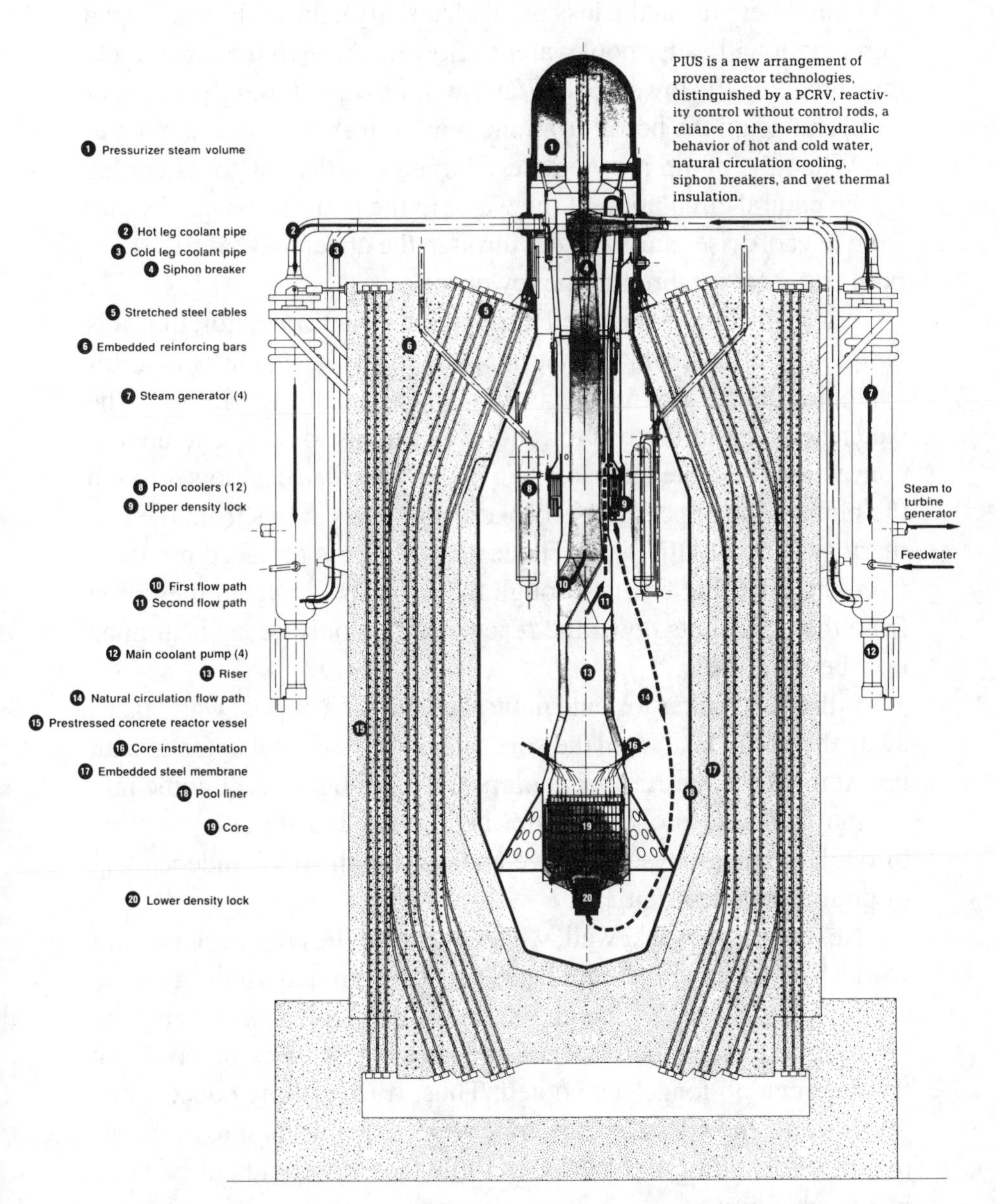

Figure 7.21 Sweden's PIUS reactor. COURTESY OF RAYTHEON ENGINEERS & CONSTRUCTORS AND ABB ATOM AB.

If the pump speed is reduced below normal, or if the pumps shut off completely (as in the loss of all electricity), the hydrostatic head of the pool will cause pool water to flow up through the lower lock, thus breaking the lower lock. Water will flow out through the upper lock, and again the boron from the pool will shut down the reactor.

When the pump is completely turned off, the reactor is cooled by the natural circulation of the water in the pool. The heated water from the core rises and goes out through the upper lock, and the cold pool water enters through the lower lock.

Normally heat is removed through the steam generator; thus this is the normal heat sink. What happens if this heat sink is lost, for example, due to loss of feedwater to the steam generators? The water heats up in the reactor and even begins to boil. The natural convection flow rate increases to a higher rate than can flow through the pumps, the upper lock is broken, and water flows from the primary system into the pool. This excess water is replaced by water from the pool that enters through the lower lock. Again the boron from the pool shuts down the reactor so that only decay heat must now be removed.

If the heat sink is lost and natural convection of pool water takes away the decay heat from the core, the pool water will begin to heat up. At this time the 12 pool coolers shown in figure 7.21 come into operation. These coolers transfer heat to air by natural convection through a radiator system above and outside the pool, independent of pumps and electricity.

Next one might ask, well, what happens if the pool coolers don't work? The water in the pool heats up and eventually boils, and the steam is lost through vents in the top of the pool. This boiling can go on for one week before enough water is lost from the pool that the core can no longer be cooled. Thus, if everything conceivable goes wrong, the operators have one week to get water into the pipes that refill the pool; it is hardly credible that this could not be done in a week's time.

What happens if a primary system water pipe breaks? Again, the water in the core boils, the density locks are broken, and water from the pool flows by natural convection through the core. This shuts the nuclear reaction down and removes the decay heat. Thus, no problem. The fuel cladding never heats up enough to allow to any fuel failures.

The speed of the reactor pump that controls the hydraulic balance always senses the level of the hot and cold water in the density locks. If the level changes, the sensors tell the pump to slow down or speed up, depending on what is needed to balance the flow.

The system does not need control rods at all. It can be completely controlled by varying the level of boron in the primary system and allowing just the right amount of water into the system from the pool.

The operation may sound tricky, but it really works for all conceivable accident situations. The Swedes have built and operated large systems that simulate the PIUS reactor using electrical heating, and they have demonstrated convincingly that the system works for normal and all off-normal conditions.

GENERATING COSTS FOR THE NEW DESIGNS

Electricity generated by the nuclear plants described in this chapter is estimated to be competitive with fossil fueled plants, though natural gas may currently have a slight advantage. The evolutionary nuclear plants hold a small economic advantage over the smaller automatically-safe LWR plants on a per kilowatt hour basis (despite the simplicity of the smaller LWRs) because of the larger power rating of the evolutionary designs and the advantage inherent in economy of size.

In 1992, the nuclear industry's U.S. Council for Energy Awareness (USCEA) published a comprehensive analysis of electricity generating costs for future plants. Their results are given in table 7.1.

The cost for oil-fired plants was much higher. A combined-cycle plant is one that burns gas. It then uses the high-temperature gas-reaction products in a gas turbine to run an electric generator; this is the first cycle. Next, for the second cycle, the lower temperature gas exiting the turbine is used to heat water to produce steam that drives a steam turbine to run a second electric generator.

Despite these cost comparisons reported by the USCEA, at the time of this writing (the mid-1990s) natural gas appears to be slightly more economical than nuclear energy in the United

TABLE 7.1 USCEA 1992 Comparison of Electricity Generation Costs

Total Power Rating for Single or Multiple Plants	*Plant Type*	*Generation Cost*
1200 MW	Evolutionary LWR	3.8 ¢/kWh
	Passive LWR	4.1 ¢/kWh
	Pulverized coal	4.6 ¢/kWh
	Gas-fired combined-cycle combustion-turbine plant	4.2 ¢/kWh
	Gasified-coal combined-cycle plant	4.8 ¢/kWh
600 MW	Passive LWR	4.5 ¢/kWh
	Pulverized coal	4.8 ¢/kWh
	Gas-fired combined-cycle combustion-turbine plant	4.3 ¢/kWh
	Gasified-coal combined-cycle plant	5.0 ¢/kWh

States. As natural gas prices rise, nuclear energy will likely eventually become less expensive than natural gas so that nuclear energy will again become the choice for base-load plants. The numbers in table 7.1 show how close the costs are for the various methods of generating electricity. The table emphasizes the importance of a reduction of even one-tenth of a cent in generating cost.

8

Disposal of Nuclear Waste

INTRODUCTORY PERSPECTIVE

Each of the economical methods of producing electricity, whether using fossil or nuclear fuels, generates waste products. Fossil fuels, by the nature of the combustion process, generate gases that contribute to the greenhouse effect (such as CO_2), often the acid-producing gases SO_2 and NO_x, and smoke. Coal, in addition, produces ash. Until recent years these products entered the environment with relatively little regulation, and no amount of regulation can eliminate CO_2 and ash. One of the remarkable assets of nuclear energy is that its waste products are entirely contained in relatively small volumes that can be completely and safely disposed, permanently isolating them from the environment. It is one of those ironies of modern life that this very advantage has been turned upside down by people who fear nuclear energy; they have led much of the public to see the generation of waste as a disadvantage for nuclear energy. They keep alive the widespread, but mistaken, notion that there is no solution to the nuclear waste problem, as though we in the nuclear energy field haven't the foggiest idea of what to do with it. We constantly hear the comment, "I think nuclear energy would be great if they could only solve the waste problem."

While it is certain that waste from fossil fuels are damaging to the environment and to public health, you can be just about as sure as anything on this earth that radiation from high-level nuclear waste generated in commercial nuclear reactors and stored in a geologic repository are never, yes never, going to cause any health problems for the public. We've been operating commercial nuclear plants and storing their high-level wastes in pools at the plants for forty years. None of this waste gets to the environment. An early experiment with reprocessing of commercial fuel at West Valley, New York, more than twenty years ago, resulted in some leakage of radioactive materials to the environment, but this is the only case of environmental contamination from high-level waste from the U.S. commercial nuclear industry,* and this had nothing to do with waste disposal. There have been cases of migration of low-level waste from low-level waste burial grounds, and this requires more vigilance in the future, but this waste problem is not the high-level waste issue that people are most concerned about. We constantly hear from the media that there are no solutions to the high-level waste storage problem—despite the fact that we have been storing it successfully for forty years.

Of course, what is meant is that there is no system in place to dispose of the waste for the thousands of years that will be required. The current federal standards, as specified by the US EPA require isolation of the waste from the environment for 10 000 years. While there are a number of methods that might be used for this long-term disposal, the one that almost surely will be used is ***geologic isolation***, meaning burial deep beneath the earth's surface.

From a ***technical*** point of view there is no need to select the best solution soon. As our experience over the last forty years has demonstrated, it is very easy to isolate these wastes from the public. We can

*I caution the reader not to confuse storage of waste from the commercial nuclear industry with storage by the U.S. defense establishment. Waste storage tanks from weapons production have leaked into the ground, but these operations were, until recently, highly unregulated and the treatment of defense waste has been totally different from the treatment of commercial waste.

continue to store the waste in pools or in dry storage casks at nuclear plant sites just as we are now doing or store them at alternate temporary sites for centuries if we so choose. Of course, we should generate the funding now for ultimate disposal later, as we are presently doing by charging for disposal 0.1 cent for every kilowatt hour of nuclear electricity each customer uses. From a ***political*** point of view, however, there is indeed a great need to hurry and establish a permanent disposal system, and that's what's really important.

A long-term method for high-level nuclear waste disposal—geologic isolation—has been selected by the U. S. Congress. A preferred site has been selected—Yucca Mountain, in Nevada—although this site must still be tested to see if it satisfies all U.S. Department of Energy (DOE) criteria. This method will be described in this chapter, together with a description of the nature of the high-level waste disposal issue. This chapter will explain why the danger to public health from effective high-level waste disposal, contrary to popular belief but consistent with the views of nearly every scientific body that has studied and reported on the issue, is virtually non-existent.

There are two kinds of nuclear waste: ***high level*** and ***low level***. High-level waste is radioactive waste that comes from the fuel of a nuclear plant after the fuel is discharged from the reactor, i.e., from spent fuel. The radiation levels from this waste are high and remain high for long periods of time. Low-level waste includes materials contaminated with radioactivity from the operation of nuclear plants, factories, hospitals, and laboratories. These include items like throw-away clothing used to protect workers in hospitals and nuclear plants, ion-exchange resins used to clean up liquids that contain small concentrations of radioactive materials, and residues from cleaning radioactively contaminated equipment like steam generators.

LOW-LEVEL WASTE

Federal laws were passed in the 1980s to establish low-level waste repositories in groups of states called ***compacts***. Until the early

1990s, nearly all low-level wastes were sent to Barnwell, South Carolina, Hanford, Washington, or Beatty, Nevada. A transition from these locations to the repositories in the compacts is supposed to occur during the 1990s. It is relatively easy to construct low-level repositories with ridiculously low environmental impact, so we will not dwell further on the low-level issue here. Research on low-level waste disposal will continue in order to learn better how to prevent movement of some radioactive materials that have a high propensity for migration through soils, and there will be debate on how expensive the repository designs have to be; but these issues will be resolved. Many in the public are unaware, however, that low-level repositories pose almost no hazard at all, so we can be sure that those who fear anything to do with radiation will do whatever they can to block the construction of even these ultra-safe low-level repositories. At the same time these same people may likely remain oblivious to the more hazardous disposal sites for ash from coal-burning power plants (which may be all right, since according to what little I have read, they too are not very dangerous).

HIGH-LEVEL WASTE

High-level waste is more challenging and more interesting;thus, the rest of this chapter will be spent on high-level wastes. There are two sources of high-level waste from fuel "burned" in a nuclear reactor. The first source is the fission products generated in the fission process. The second source is the radioactive heavy elements produced from the capture of neutrons, together with their decay products. These are widely referred to as ***actinides*** and sometimes as ***transuranics***. Actinides are the series of elements from actinium, with $Z = 89$, to lawrencium, $Z = 103$. Transuranics are those elements heavier than uranium, i.e., with $Z > 92$. The fission products are initially the more radioactive of the two groups, but we are concerned with both.

Most of the fission products decay away in less than a year and are no longer important for long-term disposal. The longer-lived fission products that are of particular interest for disposal are listed in table 8.1.

TABLE 8.1 Fission Products Requiring Long-Term Isolation

Fission Product	*Half-Life (years)*	*Activity Discharged Annually from a 1000 MW PWR Reactor (Ci/y)*
Sr-90	28	2.1×10^6
Cs-137	30	2.9×10^6
Se-79	6×10^4	11
Sn-126	1×10^5	15
Tc-99	2.1×10^5	390
Zr-93	1.5×10^6	50
Cs-135	3.0×10^6	8
Pd-107	7×10^6	3
I-129	1.7×10^7	1.0

TABLE 8.2 Most Important Actinides Requiring Long-Term Isolation

Actinide	*Half-Life (years)*
Np-237	2.1×10^6
Pu-239	2.4×10^4
Pu-240	6.6×10^3
Pu-242	3.6×10^5
Am-241	4.6×10^2
Am-243	8.0×10^3

The two fission products that have received the most notoriety are cesium-137 and strontium-90. Both have half-lives of about thirty years. It takes between 500 and 1000 years for these to decay to levels at which they are no longer of concern to the environment. Thus it is necessary to isolate high-level waste for at least this long. There are other fission products, however, that have half-lives of thousands of years, as noted in table 8.1. The most important are technetium-99, iodine-129, and cesium-135. The radioactivity associated with these isotopes is so low, however, that there is some debate regarding how long these need to be isolated.

The actinides of most concern include isotopes of neptunium, plutonium, and americium. These are listed in table 8.2. It is noted that some of these have half-lives in the tens of thousands of years. In addition, radium-225 and radium-226, resulting from long-term decay of higher actinides, become the dominant ingestion hazard in the 100 000 year range.

There are two choices regarding the form of the waste to be disposed. If the fuel were reprocessed after discharge from the reactor in order to recycle the remaining U-235 and Pu-239 in it, the fission products and a small fraction of the actinides could be fixed in a glass matrix, called borosilicate glass, and this would be the form of the high-level waste to be stored. Alternately, if we're not interested in recycling, the fuel could be stored in the form of fuel assemblies, which saves the step of reprocessing, but results in throwing away the remaining U-235 and Pu-239. For waste from commercial nuclear plants, the United States has presently made the decision not to recycle but instead to dispose of fuel assemblies directly. This was decided mostly on the basis that it is not economical to reprocess and partly from a belief (which I do not share) on the part of some policy makers that somehow this reduces the chance of weapons proliferation. U.S. defense wastes, on the other hand, have already been reprocessed, and they will be disposed of in borosilicate glass (eventually in Yucca Mountain). France, Britain, and Japan are presently reprocessing their commercial fuel so their waste is in the form of borosilicate glass.

It is interesting to consider the volume of high-level waste produced by a typical nuclear power plant. A 1000 MWe PWR nuclear plant will produce about 60 spent fuel assemblies each year. If fuel assemblies are stored, you have the volume of this many fuel assemblies, with each assembly being 13 feet tall and around 8 inches square. The final volume would actually be smaller than this since the fuel rods in the assemblies are generally reconfigured into substantially more rods per assembly after discharge from the reactor for a more economical use of storage space. If the fuel is reprocessed and the waste stored in the form of borosilicate glass, a 1000 MWe plant would produce only about 100 cubic feet of waste glass, which amounts to six glass cylinders 10 feet tall and 1.5 feet in diameter. Space in a waste disposal repository to accommodate waste from many years of operation of a large number of nuclear power plants poses little problem, even accounting for the fact that waste containers must be placed some distance apart in the repository to allow for heat generation as discussed below.

YUCCA MOUNTAIN AND GEOLOGIC DISPOSAL OF HIGH-LEVEL WASTES

The status of high-level waste disposal is discussed in this section. It is important to remind the reader that high-level waste disposal is a hot political topic, with major legislation being debated in Congress as I write. A number of both technical and political issues are still being decided. Therefore, don't be surprised if some of the concepts described here are modified further. Nevertheless, the fundamental approach to high-level waste disposal described here has widespread consensus among scientists and strong support in Congress, so I expect that most of the basic concepts will remain unchanged.

The method for disposal of high-level waste that will be used in the United States, and probably elsewhere, is geologic isolation. The waste will be placed in canisters, and the canisters will be placed in a specially engineered geologic repository deep beneath the earth's surface. The U.S. Government has selected Yucca Mountain in Nevada as the likely site for the first repository. This site will be selected if a ten-year six-billion-dollar investigation of the site's geology shows that the site is suitable for a repository. The decision for site suitability is scheduled for 1998. Fuel in such a repository is to be retrievable for a period of 100 years, after which the repository will be permanently closed.

The EPA has decreed that the waste must be isolated for 10 000 years. This duration, however, is still under debate both because scientists cannot accurately project conditions for such a long time and because the 10 000 year number appears to have little justification.

The choice of the Nevada site is the result of two federal laws passed in the 1980s. In 1982 the US Congress passed the Nuclear Waste Policy Act of 1982, which established a national policy for high-level waste disposal, including a timetable and a funding policy. While this act called for the selection of three sites to be studied, or "characterized," for suitability for a geologic repository, the act was amended in 1987 when Congress passed the Nuclear Waste Policy Amendments Act of 1987. This act directed the DOE to focus solely on the Yucca Mountain site for site characterization.

A photograph of Yucca Mountain appears in figure 8.1. Yucca Mountain, named for a cactus-like bush called yucca, is located in southern Nevada, as shown in figure 8.2. It is 90 miles (145 km) northwest of Las Vegas and just at the western boundary of the Nevada (nuclear weapons) Test Site. The Yucca Mountain site is 20 miles east of Beatty, 15 miles north of Amargosa Valley, and 30 miles north of the Devil's Hole National Monument, all three of which are in the Amargosa Desert. The Amargosa Desert is a closed drainage basin from which water does not escape. The nearest extinct volcano is in the Amargosa Desert and is a "cinder cone" called Lathrop Wells. Across the Funeral Mountains in the Amargosa Mountain

Figure 8.1 Yucca Mountain. COURTESY OF DOE.

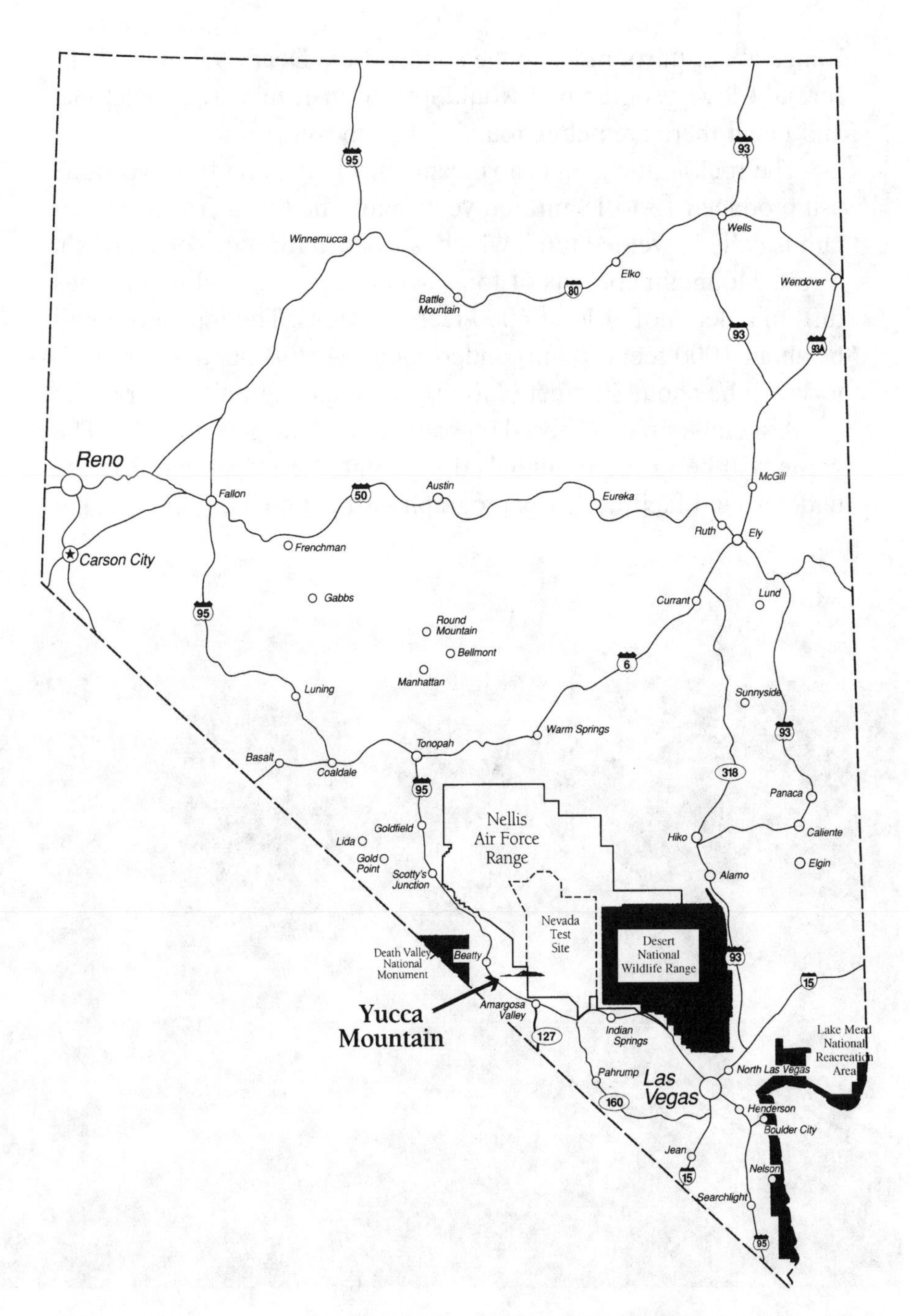

Figure 8.2 Location of the Yucca Mountain site in Nevada. COURTESY OF DOE.

Range about thirty miles to the southwest is Death Valley, in California. I love geography. I would love to visit this area sometime, and I hear there are public tours of the repository site.

The rock in the area is a volcanic tuff that came from volcanic ash produced 13 to 18 million years ago. The tuff at Yucca Mountain is called "welded tuff," which is a dense form of volcanic ash. Yucca Mountain consists of four layers of porous and non porous tuff, to a depth of at least 6000 feet (1800 m). The repository will be about 1000 feet (300 m) underground. At this depth the repository will be about 800 feet (240 m) ***above*** the region's water table.

A sketch of the proposed repository is shown in figure 8.3. The waste will be stored in tunnels drilled into the volcanic tuff in the underground facilities. A photograph of the tunnel currently being

Figure 8.3 Schematic of a geologic repository at the Yucca Mountain site. COURTESY OF NUCLEAR ENERGY INSTITUTE.

drilled into Yucca Mountain is shown in figure 8.4. Experiments are being conducted along the tunnel to determine the suitability of Yucca Mountain for permanent waste disposal. This tunnel will also be used in the actual repository to transport waste canisters to their permanent disposal location. The spacing of the canisters will be governed by the heat generation rate in the waste and a specified limit for the temperature of the rock in the repository.

Figure 8.4 Tunnel being drilled into Yucca Mountain. COURTESY OF DOE.

MULTIPURPOSE CANISTER SYSTEM AND FUEL EMPLACEMENT IN THE REPOSITORY

Spent fuel assemblies will be shipped to the repository in special containers. The exact methods for handling spent fuel have yet to be definitely decided, but it appears that DOE will adopt a ***multipurpose canister system*** for this purpose. Operation of this system is illustrated in figure 8.5. After several years in a spent fuel storage pool, the fission product heating will have decayed to a level where the fuel can be easily cooled by natural convection in air. The spent fuel will then be removed from the spent fuel storage pool and placed in a sealed thick-walled metal container, referred to as a ***multipurpose canister*** (MPC). The canister will then be welded shut, never to be reopened. Next, the canister will be placed in a ***cask***, which is a heavier container designed to remain intact even in severe accidents and to shield workers from radiation from the spent fuel.

Separate casks will be built for storage, transportation, and permanent disposal. The MPC will first be placed in the ***storage cask***

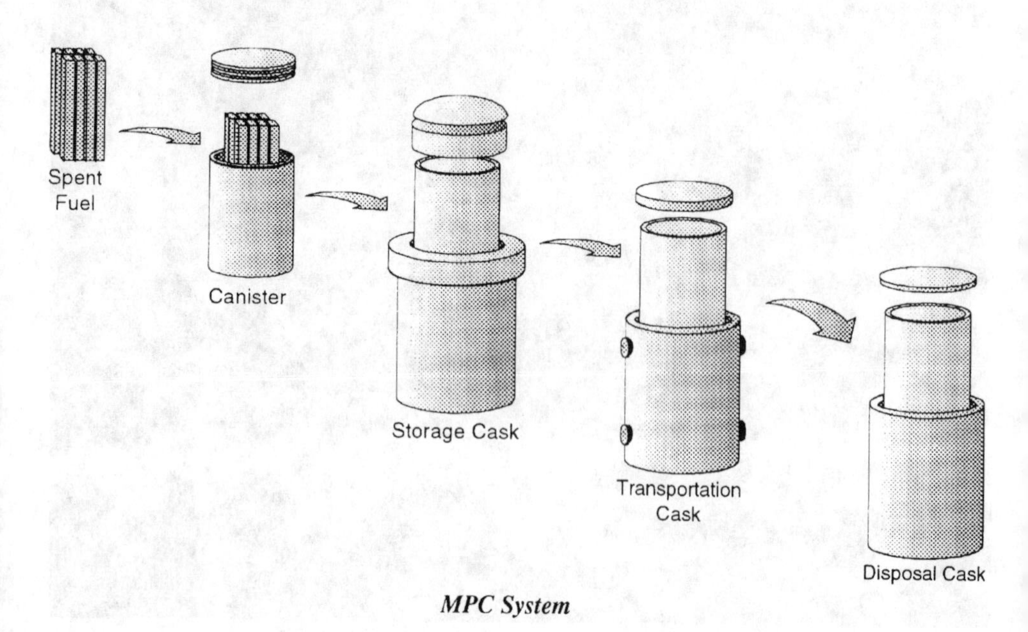

Figure 8.5 The multipurpose canister system. COURTESY OF DOE.

for ***dry storage*** at the nuclear plant site. From here on the spent fuel will be cooled by natural convection in air. When the fuel is to be shipped to the repository, the MPC will be transferred to the ***transportation cask***, specially designed for this purpose. A schematic design for the transportation system is illustrated in figure 8.6. Most canisters and casks will be massive enough that they will be shipped on specially designed rail cars and by dedicated trains carrying only spent fuel. The MPC/fuel/trasportation casks are to be made in two sized, 1 weighing 125 tons and 1 weighing 75 tons. At a few plants that do not have rail access, canisters and casks small enough to be loaded on trucks might be used, or heavier casks might be loaded on barges or heavy-haul trucks and transported to the nearest railroad.

At the repository, the MPC will be transferred again, this time to the ***disposal cask***. The cask containing the MPC will finally be emplaced in a tunnel at the repository in either a vertical or a horizontal position, as shown in figure 8.7. Prior to final closure of the

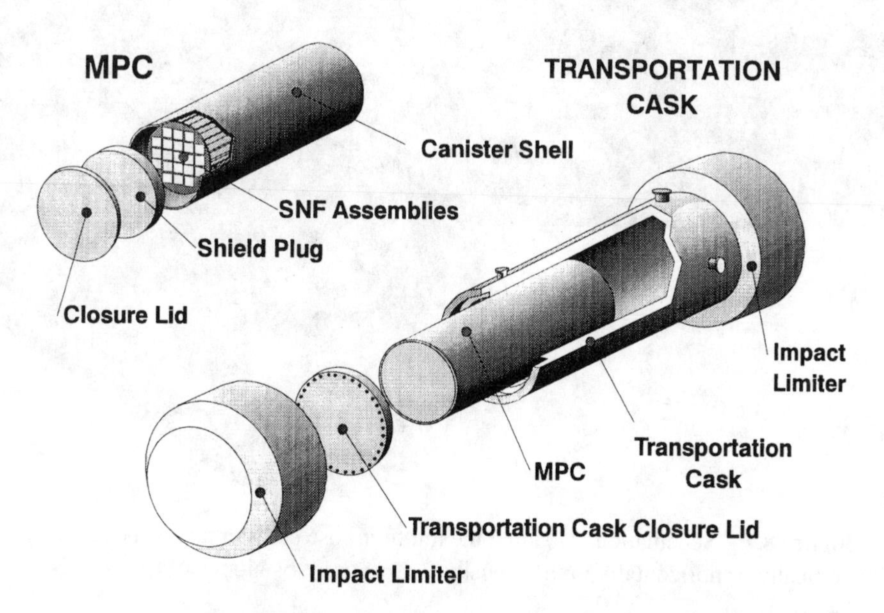

Figure 8.6 Schematic of a multipurpose canister in a transportation cask. The canister shown can hold 21 spent fuel assemblies. COURTESY OF DOE.

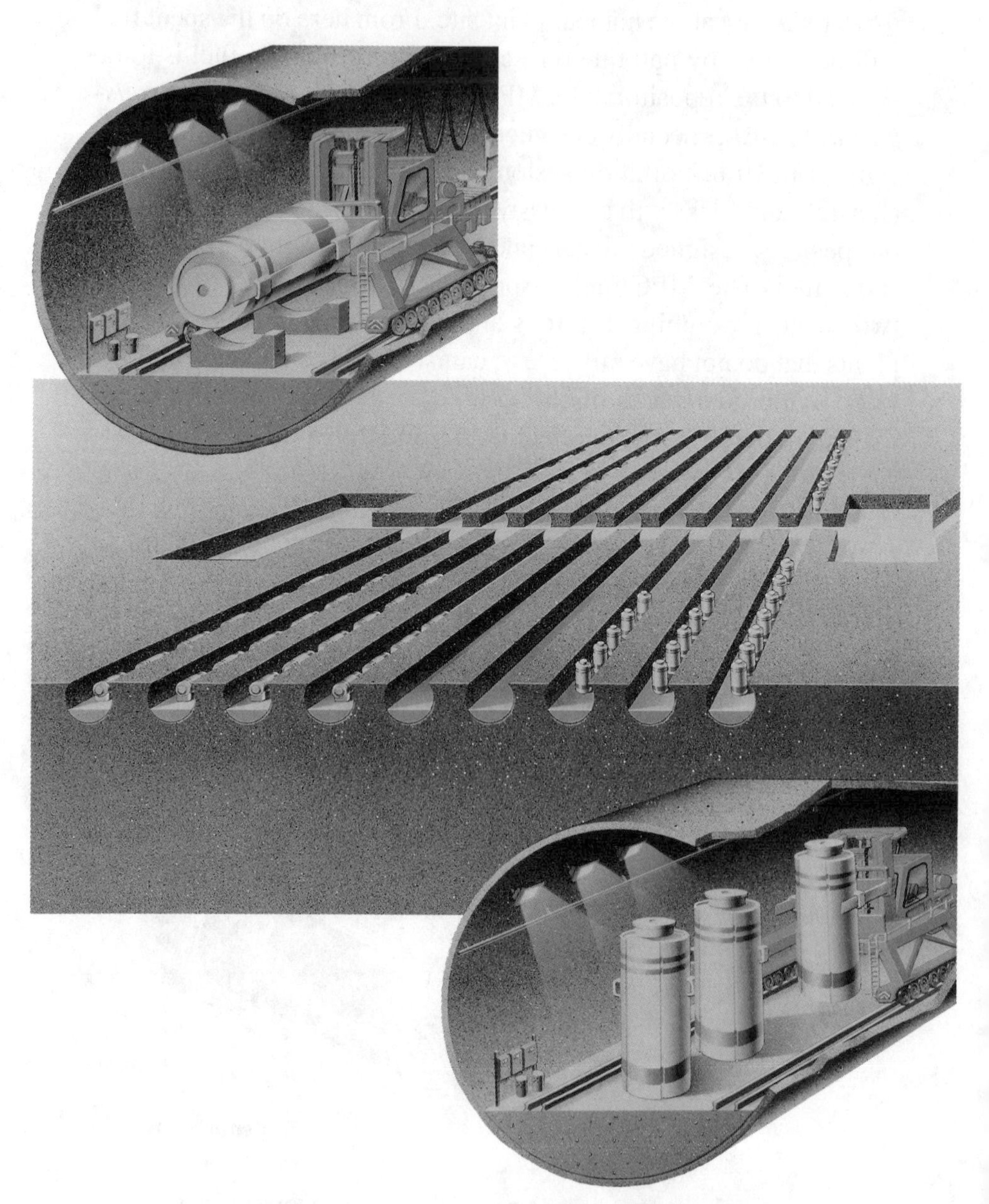

Figure 8.7 Schematic for loading the multipurpose canister and disposal cask vertically or horizontally in the repository. COURTESY OF NUCLEAR ENERGY INSTITUTE.

repository, future access to the cask and MPC would be easy if needed. Whether the tunnels would be left open or the space filled ("backfilled") with mined host rock or some other suitable material upon closure one hundred years from now has not been decided. If the space is left unfilled, for water to reach the canister, water must first partially fill the tunnel. If the space is backfilled, water must saturate the backfill material before reaching the canister. One backfill material often considered is a clay called bentonite. This clay swells when it becomes moist and then retards the further transport of water through it. Also the presence of bentonite chemically reduces the ability of water that does reach a canister to corrode the canister wall. The MPC would be sealed and would be thick enough to prevent canister failure for a very long time even if water were to reach it. More about canister standards appears later in this chapter.

Dry storage at the reactor site was referred to above as one step in the MPC system. Actually, this type of storage has already been licensed by the NRC and is currently being used at a number of nuclear plants. The dry storage facility at the Surry Power Station in Virginia is shown in figure 8.8. The casks shown weigh 117 tons fully loaded with spent fuel. They are 18 feet tall and 8 feet in diameter. The steel walls are up to 15 inches thick. The top and bottom are 12 inches thick. This design assures that the radioactive material stays inside the cask. The lid also has a double seal system, and an alarm is triggered if one of the seals fails. The weight of these fully loaded casks is about the same as the weight of the largest MPC/fuel/transportation cask being designed by DOE.

MONITORED RETRIEVABLE (INTERIM) STORAGE

In addition to a permanent repository, the Congress's 1982 Waste Disposal Act mandated that a facility for ***monitored retrievable storage*** (MRS) be considered. An MRS facility would be used for interim storage of spent fuel. At an MRS facility, spent fuel would be received in multipurpose canisters from nuclear plants throughout the country where it would be stored temporarily above ground

Figure 8.8 Above-ground dry spent fuel storage at Virginia Power's Surry Power Station. COURTESY OF VIRGINIA POWER.

in storage casks until a permanent repository is ready. Thus, storage would be part of the multipurpose canister system and would likely be similar to Surry's dry storage facility shown in figure 8.8. The fuel would be monitored at all times, and it could be removed, or retrieved, at any time. The MRS would occupy several hundred acres, with a buffer zone between the stored fuel and the site perimeter. The entire area would be enclosed by high-security fences monitored by a security force.

The politics of selecting an interim spent fuel storage site are tedious. Perhaps by the time you read this an interim storage site will have been selected. For several years the US government tried to find states and Indian tribes to volunteer a site that would then be examined for suitability. The groups volunteering were well compensated for even allowing their sites to be studied. Early in this process several counties and Indian tribes volunteered sites for consideration. Private development of a site is underway by an Indian tribe in New Mexico. At the time of this writing, DOE sites near Yucca Mountain in Nevada, and elsewhere, are being considered by the U.S. government. Despite public fears, for those of us who understand radiation and nuclear fuel storage, it is difficult to imagine a safer and cleaner industrial operation than an MRS facility.

SPENT FUEL SHIPPING

At the nuclear plant, spent fuel, already in multipurpose canisters, will be loaded into heavily shielded transportation casks, as illustrated in figure 8.6. These casks will be shipped to the MRS facility or to the repository by rail. Tests have been performed on transportation casks that involve casks being struck by high-speed trains, trucks crashing into solid barriers at high speed, casks being subjected to fires, submerging in water, and falling from several feet onto a steel spike. Cask designs have to be leakproof after these tests in order to qualify for spent fuel shipping.

Schematic illustrations of the tests that shipping casks must pass are shown in figure 8.9. Some pictures of a test in which a truck carrying a transportation cask hit a solid wall at more than 60 miles per hour are shown in figure 8.10, and none of the simulated radioactive material inside the cask leaked out.

HOW MIGHT WASTE GET FROM THE REPOSITORY TO THE ENVIRONMENT?

Now let us examine the extent to which a repository at Yucca Mountain can safeguard the environment. The primary pathway for nuclear waste to reach the environment from a repository is by transport with water. Therefore, it is useful to examine the water situation at Yucca

How are casks tested?

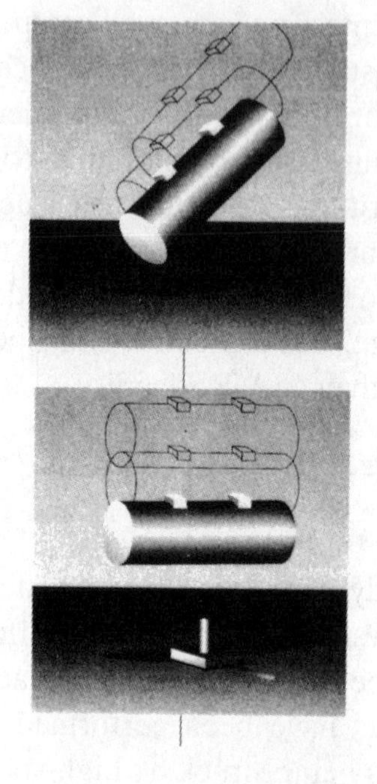

Free Drop — The cask is dropped from 30 feet onto a flat, unyielding, horizontal surface so that the cask strikes its weakest point.

Puncture — The cask is dropped from 40 inches onto a steel bar 8 inches high and 6 inches in diameter at a point where damage is most likely to occur.

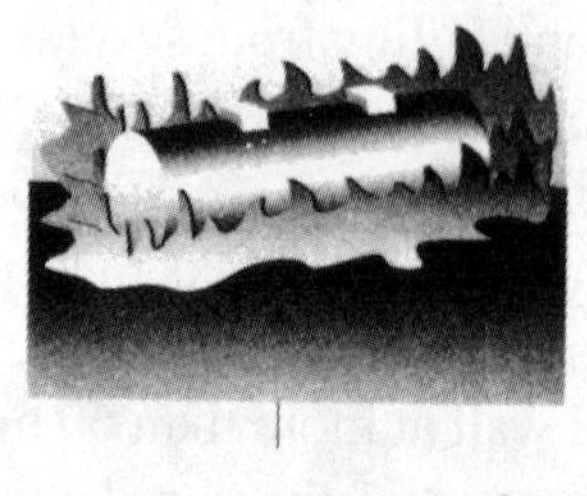

Thermal — The entire cask is kept for 30 minutes in a jet fuel fire burning at a temperature of 1,475°F.

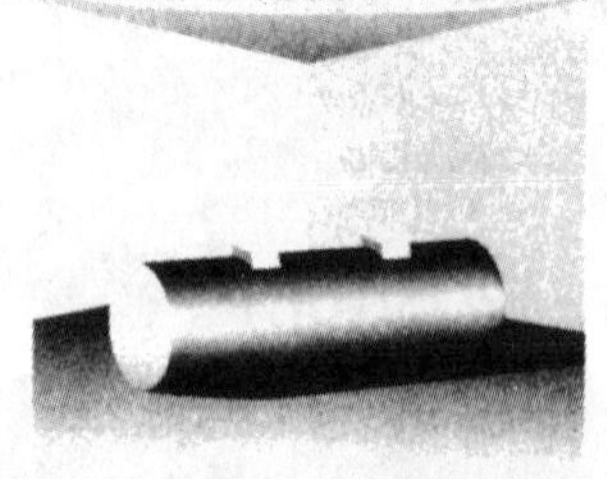

Water Immersion — The cask is totally immersed under 3 feet of water for at least 8 hours.

— In a separate test, another cask is tested below 50 feet of water for at least 8 hours.

Tests of spent fuel casks show that they can withstand severe accident conditions.

Figure 8.9 How spent fuel casks are tested. COURTESY OF DOE.

Figure 8.10 Test of a spent fuel cask on a truck hitting a solid barrier at more than 60 mph, which resulted in no leakage from the cask. COURTESY OF DOE.

Mountain. A schematic of the geography and repository location is shown in figure 8.11.

Annual rainfall at Yucca Mountain is about 6 inches (15 cm), most of which evaporates. It is expected that only an extremely small fraction of that rain, if any, would reach the area where the repository will be located. The repository itself will be located above the water table in an ***unsaturated zone*** where there is relatively little water in the rock and in which water moves very slowly.

The nearest surface water into which any potentially contaminated ground water from Yucca Mountain could discharge is about 3 miles (5 km) away in the Amargosa Desert. Groundwater flow is not like a stream but rather a dampness that very slowly migrates through ***saturated*** rock. The rate of flow is estimated to be of the order of a foot per year! At this rate it would take between 10 000 and 20 000 years for groundwater to reach the Amargosa Desert from Yucca Mountain! Yes, I mean 10 to 20 thousand years. That's no typo. Of course we do not know exactly what this rate of flow is yet. The ten-year characterization of the region will provide us with a more definitive answer, so keep tuned.

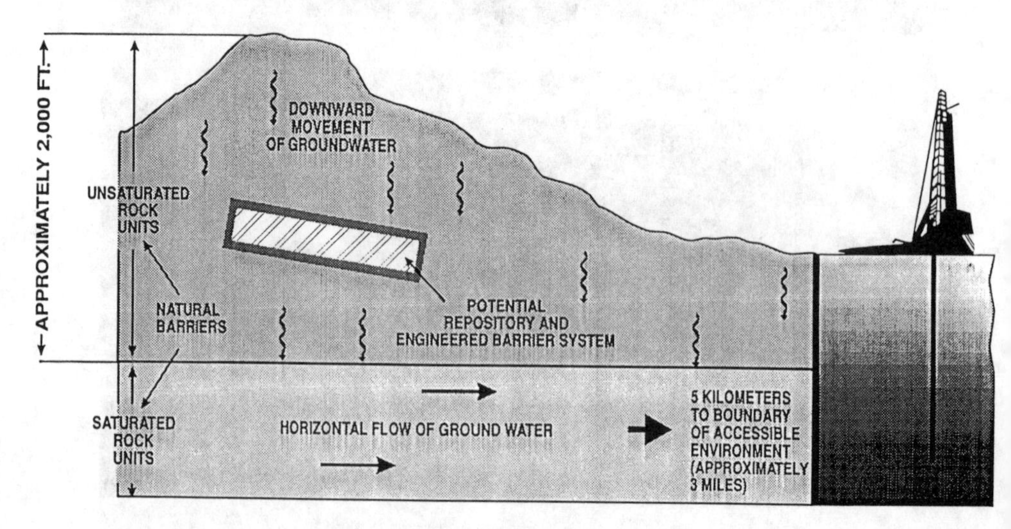

Figure 8.11 Potential water flow paths at Yucca Mountain, showing unsaturated zone and saturated (groundwater) zone. COURTESY OF DOE.

Geologists tell us that there can never be a guarantee that geology and climate will not change in unpredictable ways over the 10 000 year time period specified for isolation by current federal standards. Thus, one must investigate the question of what happens if water does reach the canisters in the repository in some significant amount. First, if bentonite clay is used as a backfill material, the presence of this clay would retard corrosion of the canister wall. Second, there is the selection of canister wall material. Current candidates for the container material for the U.S. canister design are stainless steel, carbon steel, and copper-based alloys. The federal standards require that the DOE demonstrates that the canisters can contain the radioactive material for 300 to 1000 years. (This can be done, by the way, by "accelerated testing"; the DOE doesn't really have to wait for hundreds of years to provide this demonstration!) The Swedish canister designs specify the use of 4 inch (10 cm) thick copper walls that would require 1 million years for water to corrode through, even in the presence of oxygen and sulfides in the rocks. Of course, research continues on the effect of defects like pitting on corrosion rates that may alter containment times. The French currently prefer titanium for the canister walls. The Canadians are considering titanium-based alloys and copper. The Swiss have a 10 inch thick steel canister design to provide 1000 years of isolation. Next there is the matrix in which the waste is fixed. This is the ceramic uranium oxide if fuel assemblies are disposed, as is the U.S. plan. The matrix would be borosilicate glass if the fuel is reprocessed before disposal. Several thousands of years would be required to dissolve the uranium oxide fuel and its waste. More than a hundred thousand years would be required to dissolve the waste in borosilicate glass.

Next, the waste must be transported through the rock. First it must migrate through the unsaturated rock to the groundwater. The volcanic tuff at Yucca Mountain contains a type of mineral called "zeolites," which are very effective at removing other minerals from water. As water containing dissolved waste materials migrates through the rock, some of these materials would stick to the zeolites in "ion exchange," "adsorption," and other physical removal processes, thereby slowing the net movement of the waste through the rock. These are the same processes used to soften water, and in fact zeolites are often used for this purpose. An artist's conception of the interaction of zeolites and radionuclides is shown in Figure 8.12. In the drawing, moisture is shown moving through tuff that contains zeolites. As the moisture contacts the zeolites, some of the radionuclides are attracted to and stick to the zeolites. In addition, some of the radionuclides contact and stick to the tuff.

Once the water-bearing waste reaches the water table, the water will slowly move to the environment. Again, removal processes like those described for the zeolites would occur during this movement so that the net rate of flow for most of the waste would be far slower than the already very slow flow of the groundwater.

These removal processes are complicated, however. More will be said about this in a later section, especially about which wastes are more and which are least affected by these removal processes. One possibility for speeding up transport by water is transport along earthquake fault lines. This mode of transport will be explored extensively in the characterization of Yucca Mountain, and disposal sites will be kept away from fault lines. Tritium from nuclear weapons tests has been found over 1000 feet below the surface in a fault line near the proposed Yucca Mountain repository site.

In summary, it is expected that if Yucca Mountain proves to be a satisfactory site for a repository based on the ten-year characterization study, little water will ever get to the disposal containers. If it does, the containers and waste matrix will prevent the loss of waste for thousands of years. Even if some is lost, then the flow through the rock will prevent the waste from reaching the environment for more thousands of years.

Ah! But what about earthquakes and volcanoes? These too will be studied in the site characterization. Some things are known already, however, from studies at the site over the past ten years.

Yucca Mountain has existed in its present form for at least a few million years, despite the occurrence of many earthquakes. Experience with earthquakes throughout the world has demonstrated time and again that underground structures such as tunnels, mines, and

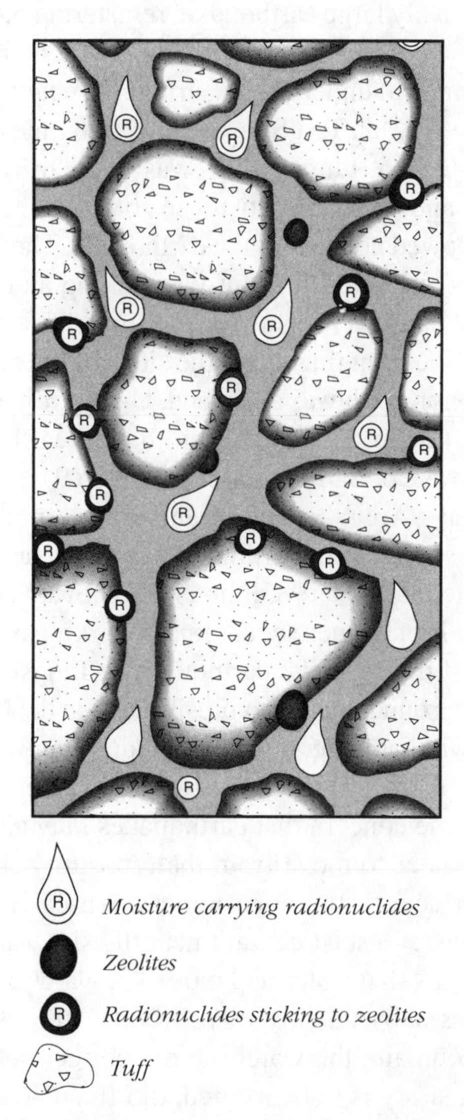

Figure 8.12 Schematic drawing of moisture flow through volcanic tuff containing zeolites, as in the unsaturated rock in Yucca Mountain, showing removal of radionuclides by the zeolites and tuff.

wells can withstand earthquakes far better than structures above ground. A relatively large earthquake, registering 5.6 on the Richter scale, occurred about 15 miles from Yucca Mountain in 1992, but it caused no damage to underground structures near the Yucca Mountain site. At the Nevada Test Site, a tunnel two miles from the earthquake epicenter—a tunnel that was dug in 1983, 400 feet underground, and is similar to those that will be built at Yucca Mountain—showed no signs of the earthquake. The tunnel walls had been painted white, and the paint was flaking and peeling, but no flakes of white paint were found on the tunnel floor after the earthquake. Water pipes hanging along the sides of the tunnel were thick with dust, and none of it was disturbed. Underground workers at the Nevada Test Site did not even know that an earthquake had happened. In the wake of San Francisco's 1989 6.9 earthquake, the underground Bay Area Rapid Transit System was the only transportation system still working. After the devastating 8.1 earthquake in Mexico City in 1985, the city's subway continued to operate. Thousands of Chinese coal miners working underground near the epicenter of the 1976 7.8 earthquake at Tangshan emerged unharmed. Underground nuclear explosive tests at the Nevada Test Site have also shown that underground structures can withstand ground motion greater than that generated by earthquakes.

There is some concern that earthquakes may alter the potential pathways for water to move from the repository, 1000 feet below the earth's surface, to the environment. An early observation of water movement at a seismic fault near the site was interpreted by a Nevada scientist that water had moved upward to the surface, but this theory was subsequently disproved. In the 1992 earthquake near Yucca Mountain, the water table, 800 feet below the level at which the repository is to be located, did fluctuate, but only by 20 feet before returning to normal.

Volcanic activity has existed in the past in the Yucca Mountain area; indeed, the rock there is from volcanic ash 13 to 18 million years old. There are seven volcanic centers located 12 to 27 miles (19 to 43 km) from Yucca Mountain. The nearest is the cinder cone, Lathrop Wells, 12 miles south of Yucca Mountain. Eruptions from two of the volcanic centers may have occurred less than 10 000 years ago. Data indicate that there is no magma chamber (an underground pocket of melted rock) beneath the volcanoes near Yucca Mountain, which makes the likelihood of a new volcano in the region extremely remote. The DOE reports that, based on current knowledge, the chance of a volcano directly affecting a repository in the area has been calculated as ranging from one in 10 million to one in 10 billion per year. That's a mighty low probability even over a 10 000 year period.

In addition to other environmental issues, the ten-year characterization study of Yucca Mountain will specifically include studies of:

- How much water there is in the rock above the groundwater table, how it moves through the rock, and how the ground water moves.
- How an earthquake might affect the repository and the water table below it.
- How volcanic activity might affect the repository.

TIME SCALE FOR DISPOSAL

How long must high-level nuclear waste be isolated from the environment? A popular curve that attempts to illustrate the time needed for the decay of high-level waste is a plot of the relative "radiological toxicity" of waste as a function of time. Toxicity is related to the amount of a material that will cause a specified amount of health damage, such as a death. The relative toxicity considered here is a ratio of the toxicity of the waste from the generation of a given amount of energy relative to the toxicity of the natural uranium needed to generate this same amount of energy. The results are plotted for the two overall groups—fission products and actinides—in figure 8.13 for the case in which spent fuel assemblies are placed in the repository. The "log-log graph" used for figure 8.13 enables us to plot a very wide range of toxicities over extremely long time periods. When the relative toxicity reaches one, the waste has the same radiological toxicity as the natural uranium that was in the

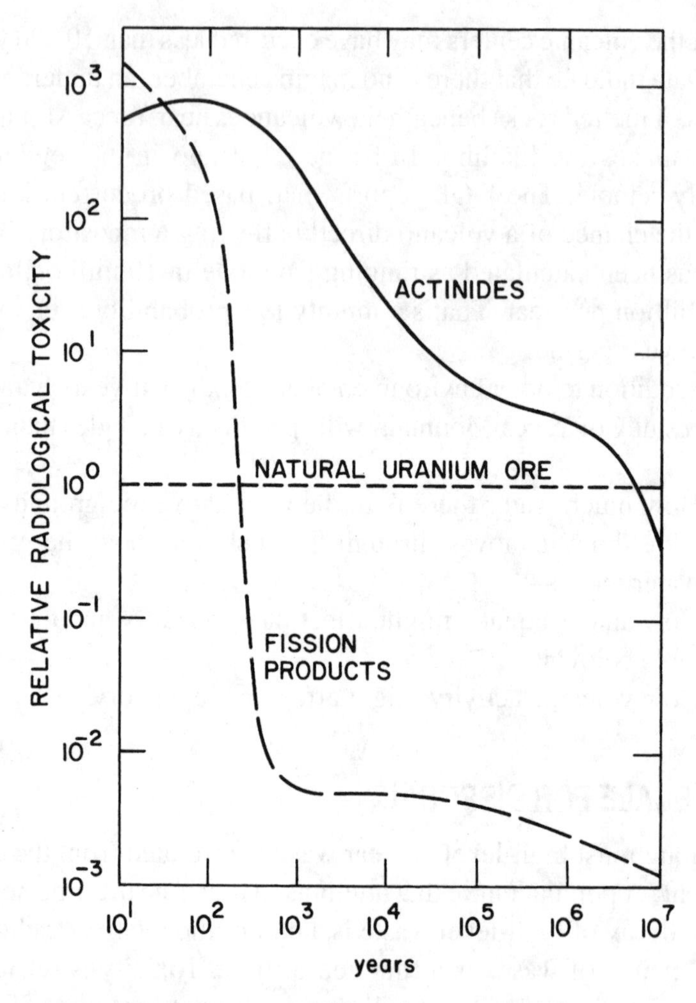

Figure 8.13 Relative radiological toxicity of high-level waste. COURTESY OF ARGONNE NATIONAL LABORATORY.

ground before it was mined. A reasonable argument is that if the waste products are no more toxic than the original uranium ore, which is located naturally in the environment, then the waste is no longer a significant danger to the public.

It is observed from this curve that the toxicity from fission products reduces to the value from the uranium ore in about 600 years.

However, the toxicity from the actinides remains high for a period even longer than 10 000 years. If the fuel were reprocessed, then only a fraction of the actinides would be included in the waste so that the time required for isolation may be reduced to the order of 1000 years.

There is a problem, however, with the above considerations based on toxicity alone. Toxicity of the waste itself is not directly a measure of risk to the public. The waste must escape to the environment before it can become a risk to people. We are ultimately interested in the radiation dose to persons who use water that might become contaminated by a repository. Thus, in addition to toxicity, risk must account for differences in the rate at which fission products and actinides can be removed from spent fuel and differences in mobility of these materials as water migrates through rock, assuming that somehow water reaches the waste. In general, fission products can be removed from spent fuel at a faster rate than actinides. Also, actinides are not very mobile; their motion would be slowed down greatly by becoming attached to the rock as the water percolates through it (as described on page 241). The same is true for all the long-lived fission products except iodine and technetium. Iodine stays completely with the water while technetium is retarded only slightly.

Professor T. H. Pigford of the University of California at Berkeley, one of the country's most knowledgeable experts on waste disposal (and another former teacher of mine), has compared the relative risks of the long-lived fission products and actinides, based on available estimates of the data needed to account for removal rates and mobility of these materials in water at the Yucca Mountain site. Values based on his results are given in table 8.3. These results are relative to what he calculates to be the most important nuclide, Tc-99.

It is observed that, for 1000 and 10 000 year transport times, the relative doses (and hence the relative risks) from the fission products Tc-99, I-129, and Cs-135 are significantly greater than from the actinides. This is so even though the toxicity of the actinides in the stored waste at the repository is higher than that of the fission products. Although these fission products may pose the

TABLE 8.3 Relative Radiation Doses for Important Spent-Fuel Radionuclides, Accounting for Hydrogeologic Transport from the Yucca Mountain Repository, based on T. H. Pigford's 1990 results.

	Relative Dose for Transport Time, t, to the Biosphere	
Isotope	*t = 1000 years*	*t = 10 000 y*
Fission Products		
Tc-99	1	1
I-129	0.5	0.5
Cs-135	0.1	0.03
Actinides		
U-234	4×10^{-6}	2×10^{-6}
U-238	6×10^{-7}	6×10^{-7}
Pu-239	1×10^{-3}	7×10^{-7}
Pu-240	2×10^{-4}	1×10^{-16}
Pu-242	2×10^{-5}	1×10^{-5}
Np-237	4×10^{-5}	4×10^{-5}
Am-243	3×10^{-5}	2×10^{-15}

greatest long-term risk from geologically isolated high-level waste, some perspective on the amount of this risk is provided in a later section.

There is a possible condition, however unlikely, in which the above argument would not be valid and in which the actinides would indeed pose the greater danger. This is for the case of a relatively sudden expulsion of radioisotopes from the repository, bypassing the long transport path through the rock, thereby allowing the actinides to escape to the biosphere without being sorbed by the rock. Perhaps such an expulsion could occur from a volcano, but the probability of a new volcano disturbing a repository at Yucca Mountain has already been defined as vanishingly small. What about an earthquake, fracturing rock enough for radionuclides to escape quickly? Although earthquakes can be expected in the region around Yucca Mountain, the mountain will not split apart, and the possibility of large faults opening up to provide pathways that would allow the rapid escape of actinides is almost certainly negligible, though the effect of earthquakes also will be studied further. That's what part of the six billion dollars to be spent on researching the area over the next ten years will be spent on. Just how remote does the possibility have to be to satisfy society? I don't have an answer for that. From what I have read so far about the Yucca Mountain site, I expect that the results of the characterization process will convince most scientists that the probability of sudden expulsion of waste from the repository will prove to be sufficiently negligible. However, it sometimes appears that for some nonscientists, just the fact that one cannot say that something is absolutely impossible is often enough to frighten them into immobility.

ACTINIDE AND FISSION PRODUCT BURNING

A completely alternative approach to high-level waste management is a process called ***actinide*** and ***fission product burning***. This requires reprocessing the spent fuel. The actinides are collected during reprocessing and are returned to a nuclear reactor, where they are fissioned, thus destroying them. From figure 8.13 it is seen that if a substantial fraction of the actinides were destroyed before disposing of the waste, the time scale for worrying about some potential sudden expulsion from the repository might be dramatically reduced. The process for destroying the long-lived fission products is called ***transmutation***. In this process, the fission product would be returned to a reactor, capture a neutron, and become another radioactive element, which would then decay rapidly. Since the Tc-99, I-129, and Cs-135 appear to be the most hazardous long-lived waste products, these, plus the other long-lived fission products listed in table 8.1, are the fission products that are candidates for transmutation.

There is a curious situation here. It is inefficient to return the actinides to a light water reactor. The actinides of concern here are the isotopes of plutonium, neptunium, americium, and curium. It is necessary to fission the actinides in order to destroy them; just capturing a neutron will merely produce another actinide. The fission cross sections of these actinides in a thermal reactor are so low that actinide burning is impractical as a waste management scheme. However, in a fast reactor, such as a liquid metal breeder reactor, the fission cross sections are high enough for the technique to be practical. As discussed elsewhere in this book, someday fast breeder reactors may be economical because they do not require high-grade uranium ore for their source of fuel. Thus they will likely be available eventually for burning actinides, both from light water reactors and from the fast reactors themselves. This ability to burn actinides and, hence, to reduce the time required for isolation of the waste is an additional incentive for the development of the liquid metal fast reactor. Also, the use of accelerators to generate neutrons for actinide burning has been proposed.

A bonus to eliminating the actinides from the waste is that the heat generated in waste that is stored a decade or more after discharge from the reactor is greatly reduced. This can result in a substantial decrease in the distance between canister storage locations in a repository and, hence, a better use of space.

It is more difficult to transmute, and thereby destroy, the long-lived fission products. Again transmutation in a thermal neutron flux is inefficient, but it is also not very efficient in a fast flux. It may be practical in an "epithermal" flux spectrum, i.e., between thermal and fast. Yet it is still not certain whether this is satisfactory. The feasibility of this way of destroying these fission products is still being investigated.

Both actinide and fission product burning will require the development of new methods of fuel reprocessing. Early results of fast-reactor metal-fuel reprocessing development are already promising in this regard. However, we are talking about issues that do not really have to be decided for years to come. There is certainly no need to burn the actinides and fission products for the first decade or two that we are using Yucca Mountain for waste disposal. If the decision is eventually made that actinide and fission product burning are in our best interest, there will be plenty of time to develop the needed methods.

HOW DANGEROUS ARE NUCLEAR WASTES?

The Very Long-Lived Fission Products

There's no question that high-level wastes must be isolated for a substantial amount of time. However, arguments above indicate that for periods beyond 1000 years, the fission products Tc-99, I-129, and Cs-135 may pose the greatest hazards to the public. So how dangerous are these materials?

The short answer is, not very. Can they ever pose a threat to the planet? There are many ways to address this question. One way is to consider the contribution of these radionuclides to the oceans' radioactivity, if in the next few tens of thousands of years, they eventually wind up in the world's oceans. While there may be little chance that water will leach these fission products from the stored fuel in Yucca Mountain, or any other equally suitable repository, an interesting question remains, what if transmutation of these radionuclides is forever impractical and a significant amount of water does reach the fuel? Then these nuclides might eventually find their way to the oceans. For this reason it is interesting to consider whether their contribution to the oceans' radioactivity is significant.

One way to approach this question is to compare the potential concentrations of these fission products in the oceans with the concentration limits for liquid effluents released to the general environment that are allowed by the NRC. A perspective on this comparison can be obtained from the following hypothetical calculation. Suppose the worldwide capacity of nuclear electricity rises to 10 000 GW, which compares to the world's 500 GW of nuclear capacity today and to the total current worldwide electrical capacity from all sources of about 5000 GW. Let us next suppose that the world has been using this much nuclear electricity for 1000 years. This example implies several things: that nuclear energy becomes and remains the preferred source for electricity; that the main reactor system is the breeder reactor; and that, if the lifetime of a power plant is 100 years, then one hundred 1000 MWe breeder reactors must be built every year! Thus this hypothetical calculation made by us mortals in the 1990s appears to be quite above a practical upper limit for what nuclear energy can supply. Our earlier assumption was that all of the three worst fission products generated over these 1000 years find their way to the ocean. Finally, we assume that the fission products are uniformly mixed throughout all the oceans, as is the case for the present impurities in the ocean. The volume of the oceans is needed for this calculation; it is 1.4 billion cubic kilometres. The fission product generation rates are also needed; the annual production rates for a 1000 MWe plant are about 400 Ci of Tc-99, 1.0 Ci of I-129, and 8 Ci of Cs-135.

The concentrations of these fission products in the ocean for our hypothetical calculation are given in table 8.4. These values are compared with the concentration limits for unregulated liquid releases to the environment that the NRC allows, called liquid effluent concentration limits. We note that Tc-99 and I-129 are factors of over 10 000 below the allowable effluent concentration limits, and that Cs-135 is still further below its limit. This shows clearly that these fission products pose no long-term threat to the ocean.

The reason the activities of these fission products are so low is a direct consequence of the fact that they have such long half-lives. A half-life of a hundred thousand years may sound scary, but the fact that they stay around so long also means that they are not very radioactive.

TABLE 8.4 Hypothetical Calculations of Concentrations in the Oceans if All Long Half-Life Fission Products Reached the Oceans

Fission Product	*Hypothetical Concentration in Oceans from 10 000 Gw for 1000 Years (μCi/ml)*	*NRC's Liquid Effluent Concentration Limit (μCi/ml)*	*Ratio: NRC Limit Divided by Conc. in Ocean*
Tc-99	3×10^{-9}	6×10^{-5}	20 000
I-129	7×10^{-12}	2×10^{-7}	30 000
Cs-135	5×10^{-11}	1×10^{-5}	200 000

The result for I-129 is especially interesting because I-129 is already present in the ocean from natural sources. The activity of this natural I-129 is 3×10^{-11} μCi/ml, which compares to the 2×10^{-7} μCi/ml effluent limit allowed by the NRC for release of this radionuclide to the environment. Thus, the concentration of natural I-129 in the ocean is a factor of 7000 below the NRC's allowable effluent limit. The most radioactive material in the ocean is potassium-40. Its radioactivity concentration is 4×10^{-7} μCi/ml, a factor of only 10 below the NRC's effluent release limit. The concentration of natural uranium-238 in the oceans is a factor of 300 below the limit; the most common radium isotope, Ra-226, is a factor of 600 below its limit. You can see that these naturally occurring radionuclides in the ocean are closer to the NRC's effluent limits than the fission products in table 8.4.

Alternate Perspective on the Danger of Nuclear Waste

Here we look at two ways to assess the danger of high-level waste that are presented by Bernard Cohen in his book ***The Nuclear Energy Option*** (Plenum Press, 1990). Dr. Cohen has a genius for this sort of thing. The first way is to compare high-level waste to other materials. The second is to examine how much money we are spending in the United States on disposing of waste, presumably to save lives of people, compared with how much we spend on other hazards.

Cohen reports his calculations for waste disposed in glass following reprocessing rather than as fuel assemblies, with only 0.5% of the actinides remaining in the waste, so the main hazard in the first 600 years comes from fission products. His primary point is the same, however, regardless of whether the waste is disposed of as glass or fuel. He first reminds us that the waste must get to people and be eaten by people before it is going to do much harm. Thus he calculates the amount of glass, somehow converted into digestible form, which would have a good chance of killing the person who eats it, and he calls this a "lethal dose." (Death in his calculation, of course, comes from cancer caused by the radiation, not from the effects of eating glass!) The following are the amounts of high-level waste glass needed to make a lethal dose, assuming typically proposed concentrations of waste in borosilicate glass:

Shortly after burial: 0.01 ounce
After 100 years: 0.1 ounce
After 600 years: 1 ounce

He then compares these with lethal doses of the following common chemicals:

Selenium compounds: 0.01 ounce
Potassium cyanide: 0.02 ounce
Arsenic trioxide: 0.1 ounce
Copper: 0.7 ounce

Thus radioactive wastes that are quite isolated from the environment are less of a hazard than a number of chemicals that exist in substantial amounts. Cohen also reminds us that arsenic trioxide is an herbicide and pesticide that is purposely scattered on ground where food is grown or sprayed directly on fruits and vegetables. It also exists as a natural mineral in the ground.

Cohen also compares the amount of high-level waste we generate each year with other materials that are dangerous if ingested. Consider the number of people who could be killed if all base-load electricity in the United States were produced by nuclear energy and if all of the waste produced annually were ingested by people immediately after burial of the waste! Pretty wild, right? Right. Further, suppose that burial occurs ten years after the fuel is removed from the reactor. Let this number be represented by the letter N. Next consider how many people could be killed if the following other substances produced annually in the United States were ingested by people. The number of people who could be killed by ingesting these substances is given below in multiples of N.

Substance	*Number of people killed by ingesting all of the following substances produced in the United States relative to N, which is the number killed by eating nuclear waste if all of the electricity in the United States were generated by nuclear power*
Chlorine gas	40 000 × N
Phosgene	2 000 × N
Ammonia	600 × N
Hydrogen cyanide	600 × N
Barium	10 × N
Arsenic trioxide	N

We don't get all upset about the fact that these chemicals are produced, of course, because they aren't eaten, or drunk, by people—plus they're very useful materials, and they've been around for a long time. It's relatively easy to keep people from eating them despite their proximity to people. What is clear is that nearly everyone is far more scared of nuclear waste, which will be buried 1000 feet below the surface of Yucca Mountain in the Nevada desert and will never be eaten or drunk by anyone, than of any of the materials listed above.

Another interesting comparison is the amount of money we spend on saving lives from nuclear waste relative to the amount we spend on other ways of saving lives. This concept will be presented more completely in chapter 9, but a few summary comments are useful here. Again I report some of the results estimated by Dr. Cohen, though calculations of money spent per life saved have been made by many authors. There are many medical tests that can save lives, e.g., Pap smears, breast cancer screening, and colon cancer tests. The costs of these are of the order of $100 000 per life saved. There are also many things that can be done to save lives on the highways, including car safety (e.g., soft dashboards, air bags, and anti-lock brakes) and highway construction (e.g., better guard rails, breakaway sign supports, and median strips). The extra cost of these measures is of the order of $200 000 per life saved. These are costs in the United States. The cost per life saved in less developed countries for food and medical care (immunizations, for example) are well known to be low per life saved, like $50 to a few hundred dollars, but then one hears the argument that these are people far away whom we don't know and for whom we cannot be responsible. Cohen compares these with the cost that the United States has decided to pay for high-level nuclear waste disposal. Based on the amount we all pay for waste disposal—0.1 cent per kilowatt hour of nuclear generated electricity—combined with Cohen's estimate of the number of deaths that would occur from the ***random*** burial of high-level waste (0.02 eventual deaths/GW·year) ***instead*** of burial at Yucca Mountain, and assuming that half of the assessed cost is to avert these deaths, he estimates the cost of the high-level waste program to be 200 million dollars per life saved! That is a thousand times the medical and highway numbers. Furthermore, the people we hope to save are people living thousands of years from now when perhaps the danger of cancer might be reduced.

Yet Another Perspective—The Natural Reactor in Africa

Scientists have discovered that about two billion years ago a natural fission reactor operated at Oklo, in the Republic of Gabon in West Africa. At that time, the uranium-235 enrichment in natural uranium was about five times the 0.7% enrichment value today. The enrichment is lower now because the half-life of U-235 is so much lower than that of U-238 that a much higher fraction of the U-235 decayed during the past two billion years than U-238. This enrichment was high enough two billion years ago that a rich uranium deposit in Gabon that was flooded with water actually went critical. That this did indeed happen was discovered when uranium ore mined in the region, to the disappointment of the mining company, was found to contain as little as 0.3% U-235 instead of the normal 0.7%. Much of the U-235 had been fissioned in the chain reaction two billion years ago until the enrichment was too low to continue to sustain criticality. Verification of the natural reactor was obtained by finding plutonium and stable fission products where the U-235 was depleted.

Another finding of interest to our discussion of high-level waste disposal is that little migration of the plutonium and fission products was observed. The natural sediment in which the reactor operated contained the reaction products naturally and quite effectively, lending further credence to the concept of geological disposal.

SOME CONCLUDING REFLECTIONS

As important as the development of a high-level waste repository is to society, it is clear that many opposed to nuclear energy will do whatever they can to block this development because "solving the waste disposal problem" would derail one of their most potent arguments against nuclear energy. No amount of characterization of Yucca Mountain will satisfy those dedicated to its failure.

As one who makes his living doing research, I know that research not only answers questions but also always raises many new ones. That is the very nature of scientific inquiry. After six billion dollars or more of research is spent on characterizing Yucca Mountain, that area will probably be understood as thoroughly as any place on Earth. At the same time, however, surprises in the geology will be encountered, and I shudder to think of how many questions six billion dollars of research can raise. No matter how well the area is characterized, some scientists will want more research to seek the hopeless goal of eliminating all uncertainty, and you can be sure that opponents of nuclear energy will exploit all unanswered questions through legal means and media hype to block or at least delay the repository. Unfortunately, the only place you'll ever find absolute certainty is at the end of the rainbow.

In addition to these inevitable questions, there are two more things that opponents are desperately hoping for—the discovery of either an endangered species in the area or useful mineral resources beneath the mountain. Either could scuttle the project.

Even if Yucca Mountain turns out to satisfy DOE's criteria for a repository, the final decision to build the repository will require extraordinary leadership from DOE.

As a final note, I would summarize the high-level waste situation as follows. Disposal of high-level waste in a repository like Yucca Mountain is extremely safe because important fission products will decay away in less than 1000 years, it is virtually impossible for enough actinides to reach the public to harm anyone, and the radioactivity of the long-lived fission products is so low that even their improbable escape to the environment will not endanger the public.

9

Comparison of Risks

In a recent headline article on the front page of my local (Albemarle County) newspaper, it was announced in the opening paragraph, "Every day more than 100 trucks, railroad cars and other vehicles containing hazardous waste pass through Albemarle County. They carry substances ranging from gasoline to deadly radioactive waste." The list then showed 649 truckloads and 28 railroad cars over a 5 day period. One of the 677 total was radioactive waste, almost surely a shipment to South Carolina of waste from the University of Virginia—from the university's hospital, experimental labs, and nuclear research reactor—low-level waste that is mostly thoroughly innocuous but must be treated carefully according to NRC regulations. When questioned by one of the professors in the University of Virginia Nuclear Engineering Department, the author of the article said that he had not used the adjective "deadly" in describing the radioactive waste in the opening paragraph; it was inserted by the editor.

This incident illustrates the way that the media take advantage of the fear of radiation among the public. The range of wastes presented gasoline as not very dangerous, since we are so used to seeing gasoline trucks on the highway, to the most "dangerous" of all—radioactive materials. It is ironic that two days after the article appeared, a gasoline truck in New York crashed on an interstate highway and five people died in the resulting fire. It is also ironic that many deaths occur every year from non-nuclear energy related accidents, but the energy source that people fear most is still nuclear.

In this chapter I diverge from the more exact numbers that are given in the rest of the book and instead talk about relative orders of magnitudes. The numbers are approximate and are intended to provide a perspective on relative risks from the various ways of generating electricity.

ENVIRONMENTAL BENEFITS OF NUCLEAR ENERGY

People in the nuclear energy field are forever perplexed by the view held by so much of society that nuclear energy is dangerous to the environment. The view held by nuclear energy professionals is quite the opposite—nuclear energy has the least negative impact on the environment of any of the ways of generating electricity. Some of nuclear energy's positive features include:

- Almost no release of any toxic substances at the power plant.
- No gases that contribute to the greenhouse effect (greenhouse gases) are released.
- No SO_2 or NO_x gases are released to make acid rain.
- The amount of ore mined per unit of electric energy generated is small.
- Only small quantities of fuel need to be transported.
- The waste volume is small and its disposal is completely controlled.
- The safety record regarding health damage to the general public (everywhere outside the former Soviet Union) is superb.

THE GREENHOUSE EFFECT

Nuclear energy's main competitors for electricity generation are the fossil fuels—coal, gas, and oil. All fossil fuels produce greenhouse gases when they burn. Nuclear energy and the renewable energy sources do not. The reader should understand something about the greenhouse effect before going on since this is clearly such an important advantage of nuclear energy.

Radiant energy comes to the earth from the sun. Radiant energy is electromagnetic, like light, and is in the form of waves, with wavelengths and frequencies. All surfaces emit radiant energy, also called thermal radiation, a form of heat. Everyone is aware of the thermal radiation emitted by a red hot poker. Many are unaware, however, that even at room (or ambient) temperature, everything is constantly emitting thermal radiation. The energy emission rate is small, and you cannot see it as you can from the hot poker. The rate of radiation emission is proportional to the fourth power of the absolute temperature (i.e., proportional to T^4, where T is in kelvins). Thermal radiation is composed of a spectrum of wavelengths. The radiation from the sun is composed mostly of short wavelengths (and high frequency), including, like the hot poker, radiation in the visible range of wavelengths. Some of this radiation is absorbed by the atmosphere above the earth, but most of it is absorbed by the earth, thus warming it up. On the other hand, radiation from objects at low temperature radiate energy at long wavelengths (low frequency), in the invisible infrared region. The earth radiates much of the energy it receives from the sun, but this radiated energy is in the infrared wavelength region.

Now here is where the greenhouse effect comes in. First, let's look at a real greenhouse. Although glass is, as we all know, quite transparent to visible light, it absorbs much of the infrared thermal radiation incident upon it. Therefore, in a greenhouse, the short wavelength radiation from the sun goes right through the glass; but when the earth reradiates thermal energy at long wavelengths, the radiation is trapped by the glass. It is this trapped energy that heats up the atmosphere inside a greenhouse even on a cold winter day.

Now we can understand the greenhouse effect as it is being applied to the earth's atmosphere. In addition to air, the atmosphere contains a number of gases we will call greenhouse gases. Other than water vapor, the main one is carbon dioxide (CO_2), which comes from burning fossil fuels, decaying vegetation, and breathing. Another is methane (CH_4), which comes from natural gas, decaying debris, and animal waste. The group of fluorochlocarbons that causes such a problem with the ozone is another. Unlike air, these gases absorb much of the thermal radiation in various parts of the infrared spectrum, much like the glass in a greenhouse. Therefore, much of the thermal radiation emitted from the earth that would have escaped to outer space had the greenhouse gases not been present is absorbed by these gases. Thus an excess of greenhouse gases in the atmosphere will cause the temperature of the atmosphere to increase, in the same way as in a real greenhouse.

This temperature increase is what is called global warming. It has been well established that the concentration of CO_2 in the earth's atmosphere is increasing at a relatively high rate due to the continuing increase in the burning of fossil fuels. The extent or rate at which this will cause the earth's temperature to increase is still being debated; the problem is complex since so much can happen to CO_2; for example, it is constantly being dissolved in and reemitted from the oceans. Many scientists are concerned, however, that the problem is serious enough to have a major impact on the world's climate and to lead to the melting of enough of the polar ice caps to flood many of the world's coastal areas. (It's the ice caps that are the problem, not the icebergs floating in the ocean. Melting icebergs don't change the water level in the ocean. You can check this out by melting some ice cubes floating in a glass of water and observing whether the water level changes.)

Burning coal produces CO_2 from the combination of carbon in the coal and oxygen in the air. Burning natural gas produces CO_2. However, some of the energy from burning natural gas, which is mostly methane (CH_4), comes from the reaction of the hydrogen with air (which only forms water). Therefore, only about half as much CO_2 is produced per unit of energy from natural gas as from coal. Oil is intermediate between coal and gas. Some of the natural gas inevitably escapes from natural gas pipelines and combustion systems. The escaped methane is a greenhouse gas, but gas is still an improvement over coal.

Nuclear energy produces none of these greenhouse gases since the energy comes from fission of the nucleus rather than from a chemical reaction with oxygen. Likewise, renewable sources of energy do not produce greenhouse gases. The renewables are solar, wind, geothermal, and biomass. Biomass, which is wood or other organics grown for burning, is interesting since the energy here comes from burning carbon compounds in air, which clearly produces CO_2. The catch is that the amount of CO_2 that is released in burning the biomass is exactly equal to the amount of CO_2 that was absorbed from the atmosphere in the first place to grow the biomass. Thus burning the biomass adds no net CO_2 to the atmosphere.

RISKS

In chapter 6, I began a discussion of risks and accident probabilities. Reactor designers are tuned in to this way of comparing risks, and we constantly seek ways to put these comparisons into perspective. Let me now expand this discussion.

The technical community tries to be quantitative in its comparison of risks. This is done by defining risk to the public, R, as a function of the probability of occurrence of a hazardous event, P, and the severity of its consequences, C.

$$\text{Risk} = \text{Probability} \times \text{Consequences } (R = P \times C)$$

Thus, to the scientist, if the probability of an accident leading to a serious consequence is sufficiently low, the risk to the public may also be low, or even “negligible.” Social scientists, however, are quick to point out that risk is viewed quite differently by many members of the public. Many are especially fearful if they perceive that they do not have control over the risk and/or if they do not enter voluntarily into the risk. In such cases, many members of the public distrust the “experts” or scientific assessments of risks. One outcome of this dichotomy in perception is that often enormous sums of money are required to reduce risks further that are not very great to begin with, while sometimes relatively few resources are devoted to much greater risks. Democracies must continue to struggle with this dilemma in our technological society. In this chapter, I will be dealing with quantitative risks, as scientists and engineers believe is our responsibility to do, while acknowledging that education is only one part of the resolution to this perception dilemma.

At a recent meeting of the American Nuclear Society at the San Francisco Hilton Hotel, I sat in an audience of some 1000 engineers. Over us were several thousand large decorative glass rods hanging from the ceiling. The sudden failure of the support holding any one of these rods could easily lead to the death of someone in this large audience, many of whom make their living calculating accident probabilities and risks from accidents. And this is in earthquake country where, during the ten years I worked in California, I personally saw equipment hanging from a ceiling swing vigorously back and forth during an earthquake. I sat there in the Hilton wondering what the probability of failure was; my guess is that it far exceeded the probability that any of that audience will ever die from radiation from the nuclear industry.

Probably nobody worries about chandeliers falling—except for the characters in Peanuts. In one Peanuts comic strip, Marcie and Peppermint Patty are at a concert in which Hayden's 96th Symphony is being played. Marcie tells Peppermint Patty that in 1791, during the premier performance of that symphony, a chandelier fell on the audience. Peppermint Patty looks up, sees a chandelier above her, and asks the boy next to her if he will trade seats.

Another wonderful Peanuts comic strip involved risk. It showed Charlie Brown and Lucy sitting side by side looking up at the stars. Charlie Brown asked Lucy to tell him the probability that a meteor will strike at the very spot where Charlie Brown is sitting. Lucy answered that the probability is one chance in 1,000,000,0 . . . and the rest of the page was filled with zeroes. In the last picture Charlie Brown had moved over a little bit.

Nuclear Energy Risks in Perspective

The chance of being killed as a result of radiation from the nuclear industry is ridiculously low compared to being struck by lightening, being electrocuted, or dying from any common accident.

However, the argument goes that these are individual situations. What about the colossal accident where many die at the same time? Here many think that nuclear energy is surely the worst of all possibilities. Well, wrong again. Three Mile Island resulted in no deaths. The one nuclear power accident that resulted in early fatalities was Chernobyl; these deaths were workers putting out the fire. Nuclear critics will point to this accident despite the fact that nothing of this nature can happen in a licensed western-style commercial reactor. The number of latent cancers caused by Chernobyl can never be identified because it will be small compared to other causes of cancer in the same population. The number of identifiable deaths from Chernobyl was far below the number of people drowned in a number of dam failures. The number of identifiable deaths from Chernobyl was orders of magnitude below the thousands who lost their lives at the chemical explosion at Bhopal, India, in 1984 or the 900 who drowned on that tragic night in 1994 when a ferry sank on its journey from Finland to Sweden.

Many analyses have been published that show that the possible number of deaths from a nuclear reactor accident pale in comparison with the possible number of deaths from a liquefied natural gas explosion during its frequent transport in a ship in the harbor of one of our great seaports (from a "vapor explosion" if enough of the liquefied gas were to spill into the water), or from the failure of a dam on the American River north of Sacramento, California, or from a jet airplane from nearby O'Hare or Laguardia Airports crashing into a crowded Soldiers Field or Shea or Yankee Stadium. The probabilities of these events are very low, but so is the probability of a large accident at a western nuclear power plant. I could quote numerical estimates appearing in the literature to illustrate these ideas, but is that really useful? What is important is that, over a period of time, we erase the mistaken notion carefully nurtured by critics of nuclear energy that nuclear energy accidents pose the worst threat of all industrial endeavors. Perhaps this unfounded fear stems from a latent fear that a nuclear plant can explode like a bomb. As I discussed in chapter 6, this is an impossibility, a fact that even most nuclear critics understand, but I'm continually told that much of the public still thinks this might happen. I hope that time and the efforts of science teachers working with our youth can, over the long run, dispel such notions.

A question that troubles some people is, just how many people can be killed in the worst possible nuclear accident? It is always a dilemma for an engineer to attempt an answer to this question since the probability of a large number is so absurdly low that it is difficult to get across. Dr. Norman Rasmussen, the MIT professor who pioneered probabilistic analyses of nuclear accidents, gives an interesting answer to this question. To paraphrase his response, it goes like this.

Questioner: What is the number of people who can be killed in the worst nuclear accident that you can imagine?

Rasmussen: Let me first ask you a question. What is the largest number of people who can be killed in a plane crash?

Q: 250.

R: Then the answer to your question is five.

Q: That's ridiculous. Everyone knows that far more people than that can be killed in the worst nuclear accident.

R: But you did not answer my question correctly. What if two jumbo jets tip wings and both crash?

Q: Ok, so the maximum number is 500.

R: In that case the maximum number for a nuclear accident is 10.

Q: That's just as ridiculous as your first answer.

R: But you still did not answer my question. What if the two planes tip wings over LaGuardia Airport and one crashes into the crowded stands at Yankee Stadium and the other crashes into the crowded stands at Shea Stadium?

Q: Come off it, Dr. Rasmussen. That scenario is utterly ridiculous; as a practical matter, that's impossible.

R: I agree, and that's also the answer to your question. The worst nuclear accident I could imagine would have an equally ridiculous probability.

A useful way to show how safe nuclear energy is has been provided by Dr. Bernard Cohen, so I again use his perspective to illustrate the point. The way to start is to compare the dangers from coal, from which we obtain over 50% of our electricity, with nuclear. There have been countless studies that conclude that generating electricity from nuclear energy is safer than from coal. In the three-year period from 1977 to 1980, one can quote the following reports that came to this conclusion: National Academy of Sciences Committee on Nuclear and Alternative Energy Systems, "Energy in Transition," 1980; American Medical Association Council on Scientific Affairs, "Health Evaluation of Energy Generating Sources," 1978; United Kingdom Health and Safety Executive, "Comparative Risks of Electricity Production Systems," 1980; State Legislature Studies for Maryland and Michigan, 1980; and U.S. Nuclear Regulatory Commission Report, "Health Effects Attributable to Coal and Nuclear Fuel Cycle Alternatives," 1977. There has never been a study that concludes that coal is safer than nuclear. Not one. Since 1980 a lot has been done to clean up coal, and the United States will benefit from the 1991 clean air legislation passed by Congress, but this will not change which is safer.

The problem with coal is the amount of carbon released in the smoke from a large coal power plant, in addition to sulfur and nitrogen oxides. With present technology such a plant releases about two million pounds of carbon per hour out the stack. It is hard to picture how much this is, but I recently heard a way that may help. Compare this amount to the number of ordinary bricks that would be equivalent in weight to this amount of carbon. The equivalent amount would be 300 000 bricks per hour! That's a lot of carbon.

A number of studies have tried to estimate the number of deaths associated with the generation of all electricity in the United States from a single fuel source by using the best probabilistic projection models available. In Cohen's first book (*Before It's too Late: A Scientist's Case for Nuclear Energy,* Plenum Press, 1983), he provides the following summary. For nuclear energy, including the entire nuclear fuel cycle, the average of the results of government-sponsored studies comes out to be about 10 per year. The number for coal averages about 10 000 per year. In Cohen's second book (*The Nuclear Energy Option: An Alternative for the 90s,* Plenum Press, 1990), he estimates the number of deaths per year from our present coal electric generating plants as 30 000 based on a 1985 Harvard University study. This high number corresponds to the conclusion that about 4% of deaths in the United States are caused by air pollution and that coal from power plants is responsible for about one-third of these. It is hoped, of course, that this number will be reduced following the 1991 law to control air pollution. There are large uncertainties in the number of respiratory deaths associated with coal, but there is no doubt that the number is very large compared with any risks from nuclear energy.

To much of the public, a "meltdown" of a nuclear reactor looms as the worst of all possible industrial accidents. The media has propagated this myth to the extent that the word meltdown is used throughout society to signify the worst of anything. For example, the severity of a large drop in the stock market is dramatized by reporting that the market has suffered a "meltdown." How really serious is this most dreaded of accidents? Its seriousness can be compared with other energy related accidents that have occurred or which are also possible.

First, what is a meltdown? If an accident occurs so that the fuel in a nuclear reactor becomes uncovered by coolant, the fuel will melt. If the fuel melts and falls to the bottom of the reactor vessel and melts through the vessel, this is called a meltdown. The fuel may, after several days, melt through the 15 feet thick concrete basemat below the reactor into the rock below where it eventually solidifies (instead of going to China, as the descriptive term "China syndrome" suggests). In the accident at Three Mile Island, a large fraction of the core did melt, and some fell into a pool of water below the core and solidified without penetrating the reactor vessel. This would not be a meltdown by the definition given above, but one could disagree with that definition and say that it was a meltdown. Thus there has never been a meltdown during the three to four decades of operation of western commercial reactors, or there has been one—at Three Mile Island—depending on your definition. Because of this uncertainty in just what a meltdown is, the nuclear safety risk analysis literature now uses the term "core damage" instead of meltdown.

For the reactors built before 1990, the calculated probability of a meltdown per year per reactor is generally about one chance in 20 000. (It will be dramatically lower for the future reactors described in chapter 7.) Most scenarios involving a meltdown terminate long before any fatalities are predicted to occur, however, so that the chance that a meltdown will occur and cause any fatality is much lower. Now there are two groups of fatalities defined in reactor safety, called "detectable fatalities" and "latent fatalities." Detectable fatalities are those caused by radiation soon after the accident, early enough to be clearly caused by the accident. Latent fatalities are more akin to life shortening. They may be cancer deaths occurring several decades after the accident, and their origin can never be proved to be radiation from the accident since the cancer rate in the same population is so much larger than the small addition that might be attributed statistically to the accident.

Using Dr. Cohen's summary of the NRC's famous Reactor Safety Study of 1975 (the "Rasmussen Report")—which we now know, based on much more elaborate analyses published by the NRC in 1990 in the report NUREG-1150 referred to in chapter 6, overestimated the dangers from nuclear accidents—the estimated detectable fatalities from a meltdown are as follows:

Number of Detectable Fatalities

98 out of 100 meltdowns	Zero
1 out of 100 meltdowns	More than 10
1 out of 500 meltdowns	More that 100
1 out of 5000 meltdowns	More than 1000
1 out of 100 000 meltdowns	3500

As you read these numbers, bear in mind that during the four decades in which nuclear energy has been used commercially in the United States, there have been no workers in the U.S. commercial nuclear industry and no members of the U.S. public who have suffered early fatalities or even radiation sickness from radiation exposure from operation of commercial nuclear plants.

What about latent fatalities? The total fatalities, both latent and early, were also estimated in the Reactor Safety Study. The estimates are:

Total Number of Fatalities

2 out of 3 meltdowns	Less than one
1 out of 5 meltdowns	More than 1000
1 out of 100 meltdowns	More than 10 000
1 out of 100 000 meltdowns	48 000

Considering both early and latent fatalities combined with the probabilities of deaths per meltdown, the average number of deaths from a meltdown, based on the Reactor Safety Study, is 400.

These numbers can be compared with the number of respiratory deaths from coal burning referred to several paragraphs back. These are mostly "undetectable deaths" that cannot be traced directly to, but can be statistically attributed to, coal burning.

With these numbers in mind, Dr. Cohen raises the interesting question, how many meltdowns must there be in the United States for nuclear energy to be as dangerous as coal? Based on the 10 000 coal related deaths per year, he calculates that we must have a meltdown every two weeks! Remember, Three Mile Island is the only meltdown in the United States in nearly 40 years of commercial reactor operation (i.e., if you consider Three Mile Island a meltdown). Now suppose we are off by a factor of 10 in our estimates of how dangerous nuclear energy is (as some critics of nuclear energy insist). Then we would have to have one meltdown every six months to be as dangerous as coal.

Risks must be treated quantitatively in order to reach logical conclusions about what is dangerous. This, of course, is what is so difficult for the public at large, which is the reason that the media can exploit irrational fears in society to sell papers and TV news shows. One of the most memorable quotes from one of the most popular books on the alleged horrors of nuclear energy (*The Careless Atom,* by Sheldon Novick, Houghton Mifflin Company, 1969, p. 105) is the following. Referring to particles of radiation, the author says:

"When one of these particles goes crashing through some material, it collides violently with atoms or molecules along the way . . . In the delicately balanced economy of the cell, this sudden disruption can be disastrous. The individual cell may die; it may recover. But if it does recover . . . after the passage of weeks, months or years, it may begin to proliferate wildly in the uncontrolled growth we call cancer."

Now that is scary. Really scary. It's why some biologists say that any amount of radiation is dangerous, no matter how small. It's particularly scary when you realize that every one of us is struck by 15 000 of these particles every second of our lives from natural radiation, or 500 billion per year! A third of these comes from potassium inside our own bodies. A medical x-ray bombards us with 100 billion of these things.

Well, wait a minute. If so many of these particles come "crashing through" our bodies and if it is so terrible, how do we survive and how come the number of people who might get cancer from natural background and medical radiation (statistically speaking) is less than 4% of all cancer cases (as discussed in chapter 2)? The answer is that only one of these particles in 30 quadrillion (that's 30 thousand trillion) causes this terrible thing to happen, and we are struck with only about 40 trillion in our lifetime. Thus our chances of surviving this horror are remarkably high.

The important message that I most want to convey by these many examples is that radiation hazards must be treated quantitatively, and not emotionally. This is the task of scientists and science teachers. We obviously have a long way to go with the general public since, for many, the use of numbers and probabilities is rarely convincing. Nearly all scientists who study radiation and nuclear energy, however, understand these numbers and translate them into a recognition of the extraordinary safety of nuclear energy. Again, it remains my hope, and my expectation, that the passage of time, without the terrible things happening that critics predict, together with the teaching of youngsters by our school's science teachers, will gradually overcome society's fear of radiation and create a general acceptance of nuclear energy.

Expenditures per Life Saved

One way to compare risks is to compare the amount of money needed to save lives. This was discussed briefly in chapter 8. This amount can be estimated for many things, so one might think that society would make the calculation and then apportion its resources according to where it would do the most good toward saving lives.

One group that makes such calculations is the Center for Risk Analysis at the Harvard University School of Public Health. In their annual report for 1991, they provide the following interesting table, reproduced here in table 9.1. They report cost per "life year" saved, which is a little more logical and useful than simply cost per life saved.

These are all the results presented in this table in their annual report. They are making the point that I am trying to emphasize. Society has been so unnecessarily afraid of radiation that they have forced nuclear power plants to spend inordinate amounts of money for safety compared with other endeavors that would have been a more effective use of available resources. Or the corollary is that control of routine radiation at nuclear plants is thoroughly safe.

Analyses of this type (and those of Dr. Cohen that will be reported next) are based on the linear, no-threshold hypothesis for cancer risk from radiation. If this hypothesis is shown to be a large overestimate of the risk (as suggested in chapter 2), the costs per life year or per life saved would be much greater than those quoted here.

TABLE 9.1 Cost Per Life Year Saved (Results from Harvard University School of Public Health)

Lifesaving Program	*Cost Per Life Year Saved (1990 dollars)*
Safety rules at underground construction sites	$52 000
Hemodialysis at a dialysis center	$56 000
Coronary artery bypass surgery	$68 000
Front seat air bags in new cars	$110 000
Dioxin effluent controls at paper mills	$5 600 000
Control of routine radiation at nuclear power plants	$165 000 000

Dr. Cohen has his results for cost per life saved too. I have gathered together many of his reported estimates, shown in table 9.2.

Repeatedly, resources spent for nuclear safety are orders of magnitudes greater than required for other safety measures that offer the same benefits. This makes the risk from nuclear energy extraordinarily low, and this is achieved at a correspondingly high cost.

TABLE 9.2 Cost Per Life Saved (Results Estimated by Dr. Bernard Cohen)

Life Saving Program	*Cost Per Life Saved (1990 dollars)*
Medical	
Fecal blood tests for colon cancer	$20 000
Mobile intensive care units (large cities)	$24 000
Mobile intensive care units (town of 40 000)	$60 000
Proctoscopic exams for colon cancer	$60 000
Pap smear	$90 000
High blood pressure treatment	$150 000
Household	
Smoke alarms	$120 000
Highway Safety	
Improved traffic signs	$31 000
Improved highway lighting	$80 000
Upgraded guardrails	$101 000
High school driver education	$180 000
Median strips	$181 000
Collapsible steering wheels and soft dashboards	$280 000
Channeled turn lanes	$290 000
Air bag	$600 000
Premium tires	$6 000 000
Large cars	$12 000 000
EPA and NRC Regulations	
EPA air pollution equipment on coal plants (scrubbers)	$1 000 000
NRC enforcement of ALARA regulations (As Low As Reasonably Achievable radiation)	$4 000 000
EPA control of radium in drinking water	$5 000 000
EPA and NRC high-level waste disposal requirements	$200 000 000
NRC reactor safety requirements	$2 500 000 000

10

A Brief Look at Fusion

Another way to obtain energy from nuclear reactions is from fusion. Fusion is quite opposite from fission. In a fusion reaction, instead of splitting a single nucleus, two nuclei are joined together to form a larger nucleus. To get useful energy out of the reaction, the two nuclei need to be light nuclei, like isotopes of hydrogen. Thus, light elements are used for fusion as opposed to the heaviest elements used for fission. It turns out that the combined mass of two light nuclei is greater than the mass of the fusion products, so that mass is again converted into energy in a fusion reaction, just as it was in a fission reaction.

Fusion occurs in the sun when normal hydrogen atoms fuse to produce the sun's energy. Fusion occurs in an uncontrolled manner in a hydrogen bomb, in which tritium (hydrogen-3) reacts with lithium-6. Of interest to us is the potential of a controlled fusion reaction in a reactor that might be used for the generation of electricity. Fusion requires extremely high temperatures, which accounts for the difficulty in commercializing this energy source. "Cold fusion" appears not to be a candidate for energy generation.

THE D-T FUSION REACTION, PLASMA CONFINEMENT, AND THE TOKAMAK

The reaction of greatest interest at the present time for a controlled fusion reactor is the deuterium-tritium, or D-T, reaction:

$$^{2}D + ^{3}T \rightarrow ^{4}He + ^{1}n \quad (17.6 \text{ MeV})$$

This reaction is illustrated in figure 10.1. The total energy generated in the reaction is 17.6 MeV, of which 14.1 is the kinetic energy of the neutron and 3.5 MeV is the kinetic energy of the helium nucleus.

The problem with fusion reactions, including the one above, is that the cross section for each reaction—which, you will recall, is related to the probability that the reaction will occur—is very low at temperatures we are used to. The cross section for fusion reactions is a strong function of temperature; it increases as the temperature rises. It is necessary to raise the temperature to millions of degrees in order for the cross section to be high enough for enough reactions to occur to make fusion practical.

Millions of degrees immediately sounds thoroughly impractical since no materials can exist as solids to contain the reacting materials at these temperatures. There is a way out—however, a very clever way. The deuterium and tritium can be made into a ***plasma***

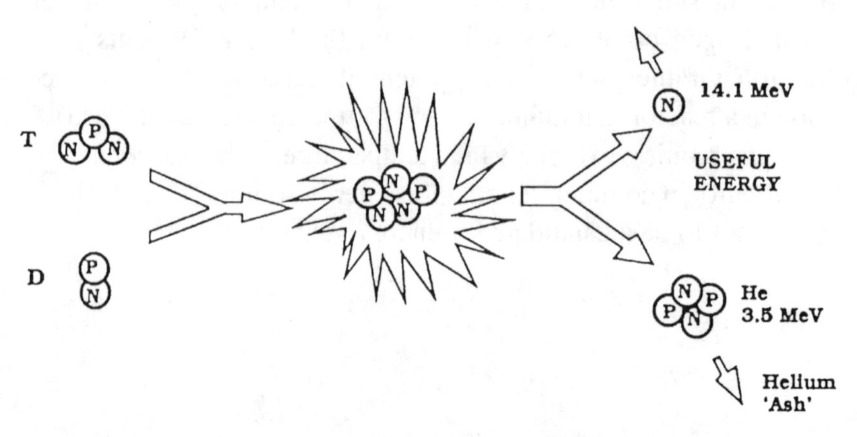

Figure 10.1 The D-T fusion reaction. COURTESY OF DOE.

that is a collection of deuterium and tritium charged ions, and these can be held together with a combination of magnetic and electrical forces without touching any solid containment walls. There are walls around the plasma, but they are not at the temperature of the plasma. The plasma is heated by the electric and magnetic fields. You see, the plasma temperature, T, is related to the kinetic energy of the deuterium and tritium ions, by the relation $E = kT$, where k is a constant called Boltzmann's constant. In turn, the kinetic energy is a function of the speed, v, of the ions since $E = 1/2\ mv^2$. The electric fields accelerate the charged ions to enormous speeds and hence to the millions of degrees required for fusion to work. Then the magnetic fields, combined with the electric fields, keep the plasma contained.

A problem with plasmas is that they are so hot that they radiate energy (by thermal radiation) at a high rate. A competition exists between raising the temperature to a high enough value to provide an acceptable reaction rate and losing energy from the plasma by radiation. The temperature of a plasma must be higher than a value called the ***ignition temperature*** in order for the reaction rate effect to overtake the energy loss effect. This temperature is equivalent to 4.4 keV for a D-T plasma. It's a sobering thought to recognize that 4.4 keV is equivalent to 50 million kelvins (or °C—at those temperatures, what's the difference?). You can check this out; $k = 1.38 \times 10^{-23}$ J/K, 1 eV = 1.60×10^{-19} J, and $E = kT$.

It is not easy to keep the plasma together long enough to raise the temperature to the necessary value. The problem is that instabilities arise so that the ions tend to escape from the plasma before reaching the required temperature. Keeping the plasma together long enough is what the physicists have been working on so hard for so long. The best device invented so far to accomplish this extremely difficult task is called the ***tokamak***, which was invented and first developed by the Soviets in the 1960s when there was a Soviet Union (and independently invented in 1963 by a New Zealander named Liley who built a prototype device in Australia).

The method of confining the plasma used in the tokamak is magnetic confinement. There is another method called inertial confinement that involves bombarding pellets of fuel with lasers or charged particle beams to compress and heat the material to ignition. This method appears not to be as promising as magnetic confinement, however, and all of the devices referred to below use magnetic confinement.

Now it may not be too difficult to get a low-density plasma up to the required temperature and hold it there for some time. It may not be too difficult to get a more dense plasma to the required temperature for an extremely brief time. However, it is desirable is to get a dense plasma to the right temperature and hold it there for a time long enough for enough of the plasma to fuse to produce power, and this is not easy. There is a famous criterion involving plasma density and confinement time, called the Lawson criterion for the British scientist who proposed it, that must be satisfied for fusion to be successful. The Lawson criterion states that the product of the fuel particle number density, n, and the confinement time for the reaction, t, must be such that, for the D-T reaction,

$$nt > 10^{20} \text{ s/m}^3 \quad \text{(D-T reaction).}$$

A schematic of a Tokamak is presented in figure 10.2. Shown are the three magnets and their orientation used to contain and accelerate the plasma up to the temperatures needed for fusion. A schematic of the Tokamak Fusion Test Reactor (TFTR) at Princeton is shown in figure 10.3, together with some dimensions and electrical parameters. A picture of the TFTR is shown in figure 10.4. A next step in the development of fusion is a large international undertaking called the International Thermonuclear Experimental Reactor (ITER).

Current fusion devices require more power input to make them work than they generate. The ratio of the power generated to input power is the equivalent gain, Q. Energy breakeven would be the case for Q = 1. Ignition, where the plasma is truly self-sustained, requires Q to be higher than one. A history of the equivalent gain for fusion experiments is shown in figure 10.5. JET is the Joint European Torus, in operation in the United Kingdom. The JT-60 is a Japanese tokamak. While TFTR and JET should reach breakeven, ignition is not expected before the operation of ITER, scheduled

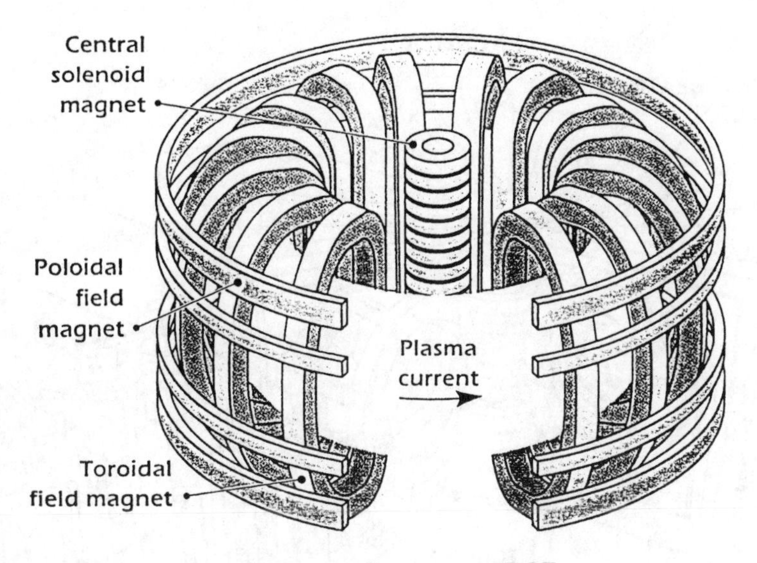

Figure 10.2 The Tokamak concept. COURTESY OF DOE.

around 2010. These devices are extremely expensive, which is one reason why development takes so long.

Not shown on figure 10.5 is a Chinese tokamak operating in Sichwan Province in the middle of the country. I had the opportunity to visit this tokamak during a reactor safety tour of China in 1985. The facility was first-rate, and the Chinese had reason to be proud of their accomplishment. The contrast between this high-tech operation and the surrounding countryside is something one often notes in developing countries; just outside the tokamak was a field being plowed by a water buffalo.

A FUSION POWER PLANT

A practical fusion power plant will need more than the plasma device. Outside the plasma there must be a way to collect the energy and convert it into electricity. There must be a way to generate tritium, assuming we're talking about a plant based on the D-T reaction. There must be a barrier between the external systems (magnets, cooling systems, etc.) and the plasma.

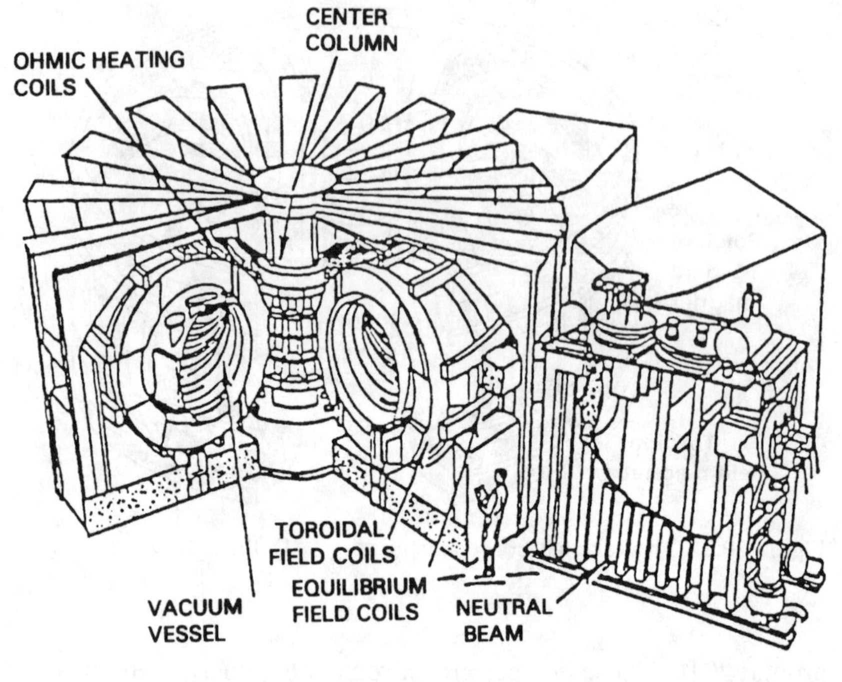

Parameters:

Major Radius	2.5 M
Minor Radius	0.85 M
Magnetic Field	5.2 T
Plasma Current	3.0 MA
Heating Neutron Beam	33.0 MW
ICRF	12.5 MW

Figure 10.3 The Tokamak Fusion Test Reactor in operation at Princeton.

Deuterium is present naturally in water so there is enough deuterium to last forever. However, tritium doesn't exist in nature; it has to be made. The way to make it is by capturing a neutron in lithium-6. Natural lithium (which contains 7.5% Li-6, with the remainder being Li-7) is placed in a "blanket" around the plasma. Neutrons produced in the D-T reaction escape from the plasma, and many are absorbed by the Li-6 to make the needed tritium. The tritium is

Figure 10.4 The Tokamak Fusion Test Reactor. COURTESY OF DOE.

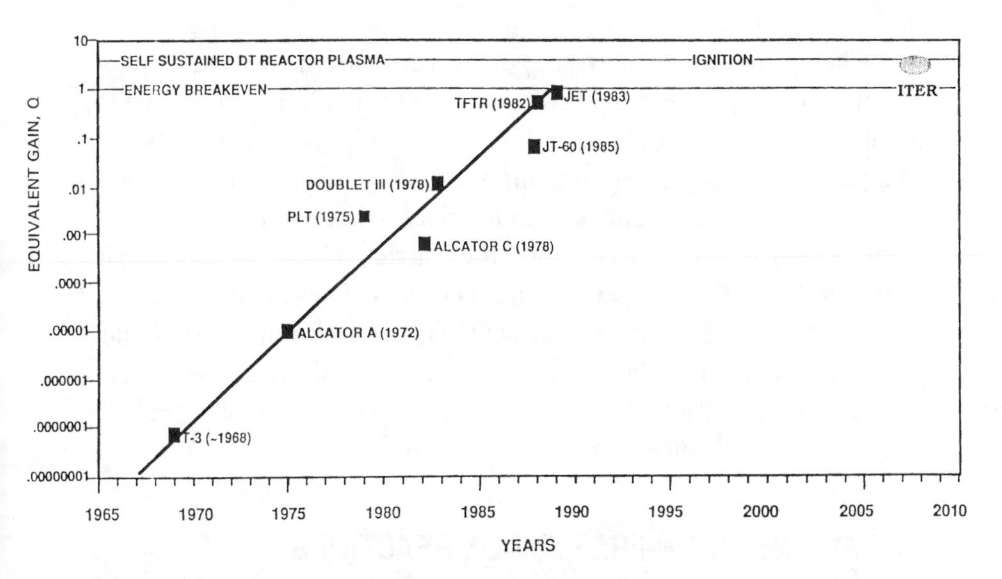

Figure 10.5 Progress in magnetic fusion. The ordinate is the equivalent gain, Q, i.e., the ratio of power generated to input power. Dates in parentheses are dates of first operation. COURTESY OF DOE.

eventually chemically separated from the lithium and recycled to the plasma. Thus, the ultimate limit to the amount of fusion power that can be produced by the D-T reaction depends on the amount of lithium in the world, not by the amount of deuterium. There is enough lithium for some time, however, so not to worry.

Another obvious question for a practical fusion reactor is, how do we convert the fusion energy into electricity? The primary way now being considered is through the lithium, which will be in the form of a liquid metal "coolant." The neutrons generated in the D-T reaction carry most of the energy—14.1 MeV of the 17.6 MeV total. These neutrons escape from the plasma and are slowed down by colliding with lithium nuclei, thus eventually giving up their kinetic energy to the coolant. The lithium is circulated through a steam generator where it transfers its heat to water to generate the steam needed to run a steam turbine to make electricity in a typical Rankine cycle. The neutrons are, therefore, the heat source used to make the steam for the turbine-generator. The 3.5 Mev from the helium nucleus formed in the D-T reaction remains in the plasma and is used to heat the plasma.

There must be a way to support and contain the external equipment around the plasma. Therefore, even though it is impossible to contain the plasma by a material barrier, a wall is necessary between the plasma and the external components, such as the lithium cooling system and the magnetic and electrical field components. In present designs, this wall is a metallic barrier. Neutrons must pass through the wall to reach the lithium coolant. These neutrons cause damage to the metal; in fact, the neutron current is much larger than is experienced anywhere in fission reactors. Wall design is a difficult challenge for a fusion plant, and it will be necessary to replace the wall at times during the life of a plant.

D-D AND NEUTRON-FREE FUSION REACTIONS

While neutrons are useful for transferring energy from the plasma to the coolant, neutrons are also quite a nuisance. Materials absorb neutrons, and this causes them to become radioactive. As long as fusion reactors are based on D-T reactions, materials are going to become radioactive, and there will be nuclear waste that must be disposed of. The idea often reported in the media that there will be no radioactive waste disposal problem with fusion is a myth, although fusion proponents suggest that metals may be made that reduce long-term radioactivity many orders of magnitude below fission reactors. Neutrons also are the source of the wall-damage problem described above. Tritium poses additional problems. Tritium is a radiological hazard, and it is not particularly easy to contain. Thus, the ideal would be a fusion reactor that produces no neutrons and does not need tritium. Someday this may actually be possible, but it's not going to happen in your lifetime.

An alternate fusion system that uses only deuterium (and hence is not limited by lithium availability), but which still produces neutrons, is based on the D-D reactions:

$$^{2}D + {}^{2}D \rightarrow {}^{3}He + {}^{1}n \quad (3.27\ MeV)$$
$$^{2}D + {}^{2}D \rightarrow {}^{3}T + {}^{1}H \quad (4.03\ MeV)$$

These reactions require higher temperatures than the D-T reaction so that the first fusion reactors will use the D-T reaction rather than the D-D reaction. Each of the two D-D reactions takes place with nearly equal probability. The tritium and the helium-3 products can react further with deuterium in the following fusion reactions:

$$^{2}D + {}^{3}T \rightarrow {}^{4}He + {}^{1}n \quad (17.6\ MeV)$$
$$^{2}D + {}^{3}He \rightarrow {}^{4}He + {}^{1}H \quad (18.3\ MeV)$$

The net result (i.e., the sum) of these four reactions is

$$6\,{}^{2}D \rightarrow 2\,{}^{4}He + 2\,{}^{1}H + 2\,{}^{1}n \quad (43.2\ MeV)$$

The added difficulty in achieving the D-D reactions is illustrated by the Lawson criterion for this system, which is

$$nt > 10^{22}\ s/m^{3} \quad (D\text{-}D\ reaction)$$

This is a factor of 100 larger than for the D-T reaction.

As stated above, it would be more desirable to eliminate neutron production entirely from the fusion reactor since neutrons generate radioactive waste. Such schemes are theoretically possible. Some neutron-free and tritium-free fusion reactions are

$$^{1}H + {}^{6}Li \rightarrow {}^{4}He + {}^{3}He$$
$$^{1}H + {}^{7}Li \rightarrow 2{}^{4}He$$
$$^{1}H + {}^{11}B \rightarrow 3{}^{4}He$$
$$^{2}D + {}^{3}He \rightarrow {}^{4}He + {}^{1}H$$
$$^{3}He + {}^{3}He \rightarrow {}^{4}He + 2{}^{1}H$$

These reactions require even higher temperatures than the D-D reactions. An obvious problem with the helium-3 reactions is the supply of helium-3. There is actually much helium-3 on the Moon, but not on Earth. It can be made artificially from a neutron capture by deuterium to form tritium, followed by tritium decay to helium-3.

Controlled fusion is a very far way off, even if its development is eventually successful. Even with the future successful operation of the international tokamak ITER, the long-term goals of the DOE fusion program are to have an operating demonstration power plant by about 2025 and an operating commercial power plant by about 2040. That's a long way off. And the history has been that every few years the projected date for commercial operation is moved back again. It is painful for me to hear people express concern about fission nuclear power with the added comment that fusion power may come along soon anyway. Often the media enthusiastically builds up hope for fusion, but don't be fooled. It probably won't happen in your lifetime, and it certainly won't happen before we have to figure out what to do about the greenhouse effect and decide how to add all the new generating capacity we will need in the next century.

11

An Even Briefer Look at Nuclear Weapons

FISSION WEAPONS

I am told by science teachers that students often ask about nuclear weapons and they, the teachers, would like to have at least a minimum of information about them. This chapter is brief for two reasons; first, I know little about weapons and, second, except for the implications regarding the commercial fuel cycle, the topic is outside the main focus of this book. The topic is mostly secret, and I have never worked in this field. However, a few ideas are widely known and can be easily described, and I can venture some opinions.

A fission weapon is made of pure metallic plutonium, nearly all Pu-239, the rest being Pu-240. The critical mass of a bare sphere of plutonium consisting of 95% Pu-239 and 5% Pu-240 is only about 10 kg and about the size of a softball. The critical mass of a plutonium sphere surrounded by reflector material would be smaller. Less than 8 kg of "weapons-grade" plutonium (to be defined later) is required for a nuclear weapon. A bomb can also be made of enriched uranium-235 (as was the bomb at Hiroshima); the critical mass of a bare sphere of uranium enriched to 93% in U-235 is 50 kg.

The trick to making a fission weapon work is to introduce a large amount of reactivity before the sphere has a chance to blow itself apart from the fission energy generated. Recall that reactivity is the difference between the criticality factor, k, and one, i.e., k-1. Once the sphere blows apart, the criticality factor falls below one, and the fission reaction is terminated. Once a large amount of reactivity is introduced and the criticality factor exceeds prompt critical (a term defined in chapter 5 under "control"), the fission rate in the weapon skyrockets so that an enormous amount of energy is generated in microseconds. If the reactivity were not introduced fast enough, the weapon would just disassemble before it generated enough destructive force and fizzle.

During the Second World War, the scientists at Los Alamos invented a method for adding reactivity fast enough. They surrounded a sphere of nearly critical plutonium with chemical explosives. Detonation of the explosives resulted in compressing the metal sphere enough to make it supercritical fast enough that it exploded with the force of an atomic bomb.

The fissions in a weapon are caused by neutrons of very high energies; we say that the neutron energy spectrum is "hard," meaning high in energy. There is no moderator to slow down the neutrons formed in fission. The spectrum is harder than that in a fast power reactor, where sodium is available for some moderation. Whereas the amount of plutonium in a weapon is less than 8 kg, the mass of plutonium in a 1000 MWe fast power reactor is around 3000 kg.

Another crucial difference between a nuclear weapon and a nuclear reactor is that it is impossible for a nuclear reactor to explode like a nuclear bomb. A nuclear reactor contains a mixture of materials, not just pure plutonium or pure uranium-235, and its geometry is entirely different from a bomb. The presence of uranium-238 in the fuel of a power reactor introduces a phenomenon called the Doppler effect, which was described in chapter 5 in the section on control. If a large positive reactivity were inserted into a reactor causing the power to increase, the Doppler effect from the excessively heated fuel would introduce a negative reactivity that would inherently and rapidly cancel the effect of the positive reactivity. An isotope equivalent to U-238 is absent from a nuclear weapon. Also, other negative feedback reactivities have time to operate in power reactors for any conceivable rate of positive reactivity addition to shut down a power excursion. Remember, to make a weapon work requires the careful placement and detonation of chemical explosives in just the right geometry and with precise, rapid timing.

I recently heard an analogy that provided insight about why a reactor cannot explode like a nuclear weapon. It goes like this. A firecracker clearly can explode. However, if you mix the explosive powder in a bucket of sand, you can no longer make it explode. So it is with the dilute fissile material in a reactor.

A plutonium weapon is made of nearly pure Pu-239, more than 90% Pu-239. Less than 10% is Pu-240. This is called ***weapons-grade plutonium.*** This plutonium is made in special reactors, not in commercial nuclear power plants. Plutonium made in commercial reactors is called ***reactor-grade plutonium***, and it is characterized by having a much larger fraction of Pu-240 than weapons-grade plutonium. There is an important reason to keep the Pu-240 fraction so low for plutonium to be weapons-grade. Plutonium-240 fissions naturally at a slow rate, but at a rate high enough to generate many neutrons. As explained above, if a bomb is to be detonated, it is important to introduce a large positive reactivity before the weapon blows apart. If a large source of neutrons is around, like that which would be produced by a large fraction of Pu-240 in the plutonium, it would be difficult to introduce a large enough reactivity to make an effective weapon; the fission rate would increase too fast, and the bomb would blow apart before the reactivity could be added. A poor bomb can be made with plutonium containing a large amount of Pu-240, but not a very effective one.

Now the relevance of all this is the following. Even though nuclear weapons can be made from reactor-grade plutonium, commercial nuclear power reactors are horrible sources of plutonium for nuclear weapons. Recall that plutonium is made in a reactor from U-238. To make plutonium with a small fraction of Pu-240, the U-238 must be irradiated for a short period of time, like a month or two. The longer it is irradiated, the more Pu-240 is made; this is because first Pu-239 is made from the U-238 and only after this can Pu-240 be made from the capture of a neutron by newly formed Pu-239. In a power reactor, fuel remains in the reactor for about four years, after which about 25% of the plutonium is Pu-240. It is thoroughly uneconomical to keep the fuel in a power reactor for only a couple of months so no power reactor would ever be so operated. Replacing the fuel every month or two is done only in reactors operated explicitly for the purpose of making weapons-grade, low Pu-240 plutonium.

The fact is that no country, other than perhaps the former Soviet Union, has ever obtained fissile material for weapons from commercial power reactors. (It is possible that the Soviet Union might have obtained weapons-grade plutonium from their RBMK 1000 reactors—the Chernobyl type—but, if so, the reactors were not being operated in a commercially economical manner.) There are too many better and simpler ways to produce weapons-grade plutonium or enriched uranium. Nevertheless, it is important that the United Nations International Atomic Energy Agency inspects the world's nuclear industry, as required by the Nonproliferation Treaty, to ensure that commercial power reactors are not suddenly being refueled every couple of months to make weapons-grade plutonium. Since reactor-grade plutonium is partially usable for nuclear explosives, however, safeguards to verify nondiversion of this plutonium are also required. Terrorists would have more difficulty making a nuclear weapon from reactor-grade plutonium than would nations. However, any group capable of making a nuclear weapon from weapons-grade plutonium must also be considered by society to be capable of making one from reactor-grade plutonium if they can obtain it.

Occasionally one reads that it is easy for a terrorist or someone else to make a bomb in a basement. In my view this is a gross exaggeration. First, all weapons that have been made so far have required substantial resources from national governments and considerable development time. Second, if one steals spent fuel to obtain plutonium, it must be shielded and handled remotely; otherwise the radiation will be lethal. Moreover, separating plutonium from spent fuel is a complex process. It is difficult to conceive of a terrorist group stealing, without detection, spent fuel assemblies, which require massive casks to transport, and even more difficult to conceive of their being able to reprocess the fuel to obtain the required plutonium. Stealing pure weapons-grade plutonium or highly enriched uranium or a bomb itself are the terrorist's best bets, although these do not involve commercial nuclear energy. (This is the reason why research reactors that used to use highly enriched (93%) uranium have switched to 20% uranium.) Stealing plutonium from commercially reprocessed fuel may be possible, but then the terrorist would have the task of making the weapon from poor, high Pu-240 content material. Isotope separation is very difficult so that enriching low enriched uranium or reducing the content of Pu-240 is clearly beyond any terrorists' capabilities. I am unimpressed by stories in the press of *students designing* nuclear weapons. Actually *building* them is the issue. Despite my view that the ease with which terrorists can make a bomb has been greatly exaggerated, people who know about nuclear weapons and terrorists' capabilities have serious concerns about terrorists making bombs, so clearly society must continue to be diligent about protecting fissile material.

FUSION WEAPONS

Fusion weapons (or thermonuclear weapons, or hydrogen bombs) operate on the principle of fusion of light elements to obtain massive energy yields. The two isotopes used are lithium-6 and tritium.

As discussed in chapter 10, the temperatures required to make fusion reactions take place are extremely high. To make the fusion weapon work, a fission bomb is first detonated within the fusion weapon to produce the high temperature required for fusion to occur. I am told that the making of a fusion weapon is a very sophisticated process indeed. Since it requires the simultaneous fabrication of a fission explosive, the making of a fusion weapon by a terrorist group would clearly be more difficult than just a fission bomb by itself.

Abbreviations

ABB	Asea Brown Baveri
ABWR	Advanced Boiling Water Reactor
ALARA	As Low as is Reasonably Achievable
ALMR	Advanced Liquid Metal Reactor
amu	Atomic mass unit
AP600	Advanced Passive 600 MWe Reactor
APWR	Advanced Pressurized Water Reactor
ATWS	Anticipated Transient without Scram
AVLIS	Atomic Vapor Laser Isotope Separation
BEIR	Committee on the Biological Effects of Ionizing Radiation
BNCT	Boron Neutron Capture Therapy
BWR	Boiling Water Reactor
CANDU	Canadian Deuterium Natural Uranium Reactor
CT	Computed Tomography
D-D	Deuterium-Deuterium nuclear fusion reaction
DOE	U.S. Department of Energy
D-T	Deuterium-Tritium nuclear fusion reaction
EBR	Experimental Breeder Reactor
ECCS	Emergency Core Cooling System
EPA	U.S. Environmental Protection Agency
EPR	European Pressurized Reactor
GDCS	Gravity Driven Core Cooling System
GDP	Gross Domestic Product
GE	General Electric

GT-MHR	Gas Turbine Modular Helium Reactor
GW	Gigawatt (one billion watts)
HPIS	High Pressure Injection System
IC	Isolation Condenser
IFR	Integral Fast Reactor
IHX	Intermediate Heat Exchanger
IRWST	In-Containment Refueling Water Storage Tank
ITER	International Thermonuclear Experimental Reactor
JET	Joint European Torus
kW	Kilowatt (1000 watts)
LET	Linear Energy Transfer
LMR	Liquid Metal Reactor
LOCA	Loss of Coolant Accident
MeV	Million electron volts
MPa	Megapascals
MPC	Multi-Purpose Canister
MRI	Magnetic Resonance Imaging
MRS	Monitored Retrievable Storage
MWe	Megawatts of electricity
MWt	Megawatts of thermal energy
NCI	National Cancer Institute
NCRP	National Council on Radiation Protection and Measurements
NPI	Nuclear Power International
NRC	U.S. Nuclear Regulatory Commission
NUREG	Nuclear Regulatory Commission report designation
PCS	Passive Containment Cooling System
PET	Positron Emission Tomography
PIUS	Process Inherent Ultimate Safe Reactor
PRHR	Passive Residual Heat Removal heat exchanger
psi	Pounds per square inch
PWR	Pressrized Water Reactor
RBMK	Former Soviet Union graphite moderated reactor design
RCCS	Reactor Cavity Cooling System

RVACS	Reactor Vessel Auxiliary Cooling System
RWST	Refueling Water Storage Tank
SBWR	Simplified Boiling Water Reactor
SEFOR	Southwest Experimental Fast Oxide Reactor
SI	Systeme International
SPECT	Single-Photon Emission Computed Tomography
TFTR	Tokamak Fusion Test Reactor
TMI	Three Mile Island
USCEA	U.S. Council for Energy Awareness
W	Watts

Index

About the Author

Albert B. Reynolds is a Professor of Nuclear Engineering at the University of Virginia, in the Department of Mechanical, Aerospace and Nuclear Engineering. He received his BS degree in Physics and ScD degree in Engineering, both at MIT. He spent nine years with the General Electric Company in California as a reactor physicist and manager in GE's nuclear energy business. In 1968 he joined the faculty of the School of Engineering and Applied Science at the University of Virginia in Charlottesville. He was formerly Chair of the Department of Nuclear Engineering and Engineering Physics at the University of Virginia. He is a coauthor with Alan E. Waltar of the book *Fast Breeder Reactors*. He is a Fellow of the American Nuclear Society.